Emerging Technologies in Africa:

Opportunities, Risks, and Pathways for Transformation

Lishan Adam

First published in the United States in 2026

ISBN: 978-1-972592-00-7 (paperback)

ISBN: 978-1-972592-01-4 (hard cover)

Foreword

Africa is at a crucial point in its development. The combination of rapid population growth, rising urbanization, and the global digital revolution creates new opportunities and challenges. Technologies—such as artificial intelligence, blockchain, renewable energy, and drones—are transforming economies, societies, and governments worldwide. The key question is whether the continent can leverage these tools to achieve its own development goals.

This publication addresses that question thoroughly, with scope and an African perspective. It examines ten emerging technologies, recognizing both their potential to drive economic transformation and their risks of widening inequality, weakening sovereignty, and centralizing power, if not properly managed.

The evidence presented here is encouraging. African innovators are already global leaders in some areas. Yet, the challenges are equally real in the foundations—infrastructure, energy, human capital, institutions, and governance. Without these foundations, even the most promising innovations will remain pilot projects rather than become drivers of structural change.

This book offers not only analysis but also practical frameworks for technology prioritization, governance design, and investment strategy. It draws on comparative experiences worldwide, not to prescribe foreign models but to highlight transferable lessons that can be tailored to African contexts.

Acknowledgements

This publication draws on my extensive fieldwork over the past twenty years across Africa and beyond—through meetings and discussions with numerous experts from government, academia, the private sector, and donor agencies, whose names cannot be listed in just a few pages. Every meeting, discussion, chat, and workshop was inspiring and enlightening. I appreciate everyone's input, time, and support.

Table of Contents

List of Tables

List of Figures

List of Abbreviations and Acronyms

3D	Three-Dimensional
4G/5G	Fourth / Fifth Generation Mobile Network
AfCFTA	African Continental Free Trade Area
AfDB	African Development Bank
AGI	Artificial General Intelligence
AI	Artificial Intelligence
AIMS	African Institute for Mathematical Sciences
AM	Additive Manufacturing
AMI	Advanced Metering Infrastructure
AR	Augmented Reality
ARSO	African Organization for Standardization
AU	African Union
AUDA-NEPAD	African Union Development Agency–New Partnership for Africa's Development
B2B	Business-to-Business
B2C	Business-to-Consumer
BPO	Business Process Outsourcing
BRT	Bus Rapid Transit
BVLOS	Beyond Visual Line of Sight
CAD	Computer-Aided Design
CBDC	Central Bank Digital Currency

CDC	Centres for Disease Control and Prevention
CERT	Computer Emergency Response Team
COMESA	Common Market for Eastern and Southern Africa
CRISPR	Clustered Regularly Interspaced Short Palindromic Repeats
CSIR	Council for Scientific and Industrial Research
DeFi	Decentralized Finance
DLT	Distributed Ledger Technology
DPA	Data Protection Authority
DPI	Digital Public Infrastructure
DRC	Democratic Republic of the Congo
EAC	East African Community
ECCAS	Economic Community of Central African States
ECOWAS	Economic Community of West African States
EdTech	Educational Technology
ESG	Environmental, Social, and Governance
EU	European Union
FDI	Foreign Direct Investment
FDM	Fused Deposition Modelling
GDP	Gross Domestic Product
GHz	Gigahertz
GIS	Geographic Information System

GPS	Global Positioning System
GPU	Graphics Processing Unit
GSMA	Global System for Mobile Communications Association
HMIS	Health Management Information System
GWh	Gigawatt-hour
ICT	Information and Communication Technology
IEA	International Energy Agency
IFC	International Finance Corporation
ILO	International Labor Organization
IoT	Internet of Things
IP	Intellectual Property
IRENA	International Renewable Energy Agency
ISO	International Standards Organization
ITU	International Telecommunication Union
IXP	Internet Exchange Point
kWh	Kilowatt-hour
LEO	Low Earth Orbit
LiDAR	Light Detection and Ranging
LLM	Large Language Model
LoRaWAN	Long Range Wide Area Network
LPWAN	Low-Power Wide-Area Network
ML	Machine Learning

MOOC	Massive Open Online Course
MW	Megawatt
NB-IoT	Narrowband Internet of Things
NFT	Non-Fungible Token
NGO	Non-Governmental Organization
NLP	Natural Language Processing
OECD	Organization for Economic Cooperation and Development
PAPSS	Pan-African Payment and Settlement System
PAYG	Pay-As-You-Go
PIDA	Programme for Infrastructure Development in Africa
PPP	Public–Private Partnership
PV	Photovoltaic
R&D	Research and Development
REC	Regional Economic Community
REIPPPP	Renewable Energy Independent Power Producer Procurement Programme
ROS	Robot Operating System
RPAS	Remotely Piloted Aircraft System
SaaS	Software as a Service
SADC	South African Development Community
SAE	Society of Automotive Engineers
SDG	Sustainable Development Goal

SLA	Stereolithography
SLS	Selective Laser Sintering
SME	Small and Medium Enterprise
STEM	Science, Technology, Engineering, and Mathematics
TB	Tuberculosis
Tbps	Terabits per Second
TVET	Technical and Vocational Education and Training
UAV	Unmanned Aerial Vehicle
UNCTAD	United Nations Conference on Trade and Development
UNESCO	United Nations Educational, Scientific and Cultural Organization
UNIDO	United Nations Industrial Development Organization
UPI	Unified Payments Interface
USD	United States Dollar
VC	Venture Capital
VR	Virtual Reality
WEF	World Economic Forum
WHO	World Health Organization
WIPO	World Intellectual Property Organization
XR	Extended Reality

PART I

Overview

Chapter 1: Purpose, Definitions and Limitations

1.1 Introduction

Africa faces a pivotal moment due to three major forces: a rapidly growing youth population; the swift progress of the Fourth Industrial Revolution, which is reshaping global industries; and urgent development needs in infrastructure, climate resilience, and inclusive growth. This combination presents both opportunities and risks.

Emerging technologies provide new solutions to long-standing development challenges in Africa and enable the region to bypass the traditional stages of industrialization seen elsewhere. However, they also introduce risks — widening inequality, displacing workers before new jobs are available, and increasing dependence on technology, which can threaten sovereignty and self-determination.

This volume analyzes the benefits and risks of emerging technologies in Africa. It is designed to serve as a rigorous analytical resource and a practical guide for the diverse actors shaping Africa's technological future.

The book has three main goals. First, it provides a comprehensive, evidence-based review of the state of emerging technologies globally and their current and potential applications in African development contexts. Second, it offers actionable analytical frameworks and strategic guidance for policymakers, researchers, entrepreneurs, and development practitioners seeking to harness emerging technologies for inclusive and sustainable development. Third, it contributes to a distinctly African discourse on technology and development—one that is informed by global experiences from Australia, Canada, Europe, East and Southeast Asia, Latin America, and elsewhere, but rooted in African realities, values, and goals.

1.2 Technology-Development Nexus

At the core of the book's analysis is the technology–development nexus: the dynamic, bidirectional relationship between technological capabilities and development outcomes. For Africa, this nexus is particularly important because technology is not merely an efficiency tool—it is a strategic instrument for structural transformation, service delivery, and global competitiveness.

This framework recognizes that technology is not just an external factor that automatically drives development outcomes. Instead, technology ecosystems (such as innovation capacity, digital infrastructure, R&D investment, and startup activity) work alongside enabling environments (including policy and regulation, human capital, financing, and institutional quality) to generate development outcomes (such as economic growth, social inclusion, climate resilience, and technological sovereignty). These results, in turn, influence both technology ecosystems and enabling environments, creating positive or negative feedback loops.

The technology-development nexus framework draws on insights from innovation systems theory (Lundvall, 2010), the capability approach to development (Sen, 1999; Nussbaum, 2011), and the technology–society co-evolution perspective (Geels, 2004). It aligns with the analytical approaches used by the OECD in its Reviews of Innovation Policy, the World Bank in its Digital Economy for Africa initiative, and the Asian Development Bank in its technology assessments for developing Asia.

This book's unique contribution is the clear integration of African development priorities—Agenda 2063, the SDGs, national plans, and principles of ubuntu, equity, and sovereignty—as both objectives and evaluation criteria.

Figure 1.1. The Technology–Development Nexus Framework

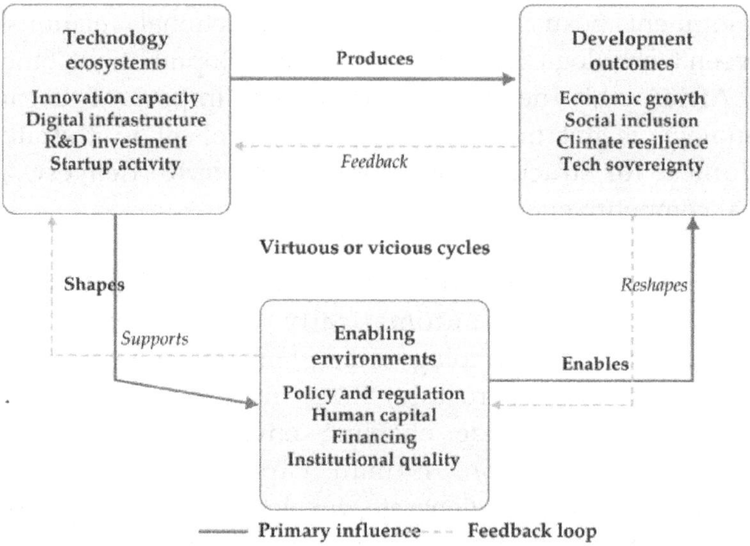

Source: Adapted from World Bank, Asian Development Bank and OECD

The book examines technology applications across sectors prioritized for their development impact, technology readiness, in alignment with Agenda 2063 and the SDGs.

Table 1.1. Key Technologies, SDG and Agenda 2063 Alignment

Sector	Key Technologies	Primary SDGs	Agenda 2063 Aspirations
Health	AI diagnostics, telemedicine, drones, genomics, IoT, robotics.	SDG 3 (Good Health), SDG 5 (Gender), SDG 10 (Reduced Inequalities).	Aspiration 1 (Prosperity), Aspiration 6 (People-Driven Development).
Agriculture	Precision agriculture, IoT, AI, drones, biotech, blockchain.	SDG 1 (No Poverty), SDG 2 (Zero Hunger), SDG 13 (Climate).	Aspiration 1 (Prosperity), Aspiration 7 (Global Partnership).
Education	EdTech platforms, AR/VR, AI tutoring, digital content.	SDG 4 (Quality Education), SDG 5 (Gender), SDG 8 (Decent Work).	Aspiration 1 (Prosperity), Aspiration 6 (People-Driven).
Financial Services	Mobile money, blockchain, AI credit scoring, InsurTech.	SDG 1 (No Poverty), SDG 8 (Decent Work), SDG 9 (Innovation).	Aspiration 1 (Prosperity), Aspiration 2 (Integration).
Energy	Solar, wind, batteries, smart grids, IoT, PAYG models.	SDG 7 (Clean Energy), SDG 13 (Climate), SDG 11 (Cities).	Aspiration 1 (Prosperity), Aspiration 7 (Global Partnership).
Manufacturing	Robotics, 3D printing, IoT, AI, digital twins.	SDG 8 (Decent Work), SDG 9 (Industry and Innovation), SDG 12 (Responsible Consumption).	Aspiration 1 (Prosperity), Aspiration 2 (Integration).
Governance	Blockchain, digital ID, AI, big data, digital payment, e-governance.	SDG 16 (Peace and Justice), SDG 10 (Reduced Inequalities).	Aspiration 3 (Good Governance), Aspiration 4 (Peace and Security).
Transport & Logistics	Drones, autonomous vehicles, IoT, platforms, AI routing.	SDG 9 (Innovation), SDG 11 (Sustainable Cities).	Aspiration 2 (Integration), Aspiration 1 (Prosperity).

Sources: African Union, 2020; World Bank, 2024; GSMA, 2023; UNESCO, 2021; IRENA; UNIDO; AfDB, 2023; ITU, 2023

Ten cross-cutting themes are tracked throughout the volume, highlighted in dedicated boxes and discussed in analytical commentary within each chapter. These themes demonstrate the interconnected nature of technology and development, ensuring that key issues are not confined to a single chapter but are addressed wherever they appear.

Table 1.2. Cross-Cutting Themes Tracked throughout the Book

Theme	Description and Relevance
Gender and Inclusion	Women in Africa are 37% less likely than men to use mobile Internet (GSMA, 2023). Gender-responsive technology strategies require addressing affordability, social norms, skills gaps, and safety. The EU's Gender Equality Strategy and India's Beti Bachao, Beti Padhao programme offer comparative approaches.
Climate and Environment	Every technology is assessed for its climate adaptation and mitigation potential, as well as its environmental footprint (energy consumption, e-waste, resource extraction). Aligned with the Paris Agreement and the AU's Green Recovery Action Plan.
Youth and Demographics	With a median age of 19, Africa's demographic dividend depends on productive employment and skills. Each technology's job-creation and displacement potential is analyzed, with implications for youth highlighted.
Inequality and Poverty	The potential of technology to reduce or amplify inequality is examined across urban–rural, income, gender, and disability dimensions. The 'leave no one behind' principle of the SDGs is operationalized.
Regional Integration	AfCFTA, AU digital strategies, and REC harmonization efforts are assessed for their role in enabling cross-border technology ecosystems and overcoming market fragmentation.
Global Partnerships	Technology transfer, South–South cooperation, development finance, and African participation in global governance (ITU, ICANN, WIPO, WTO) are tracked.
Ethics and Rights	Ubuntu-informed ethical principles guide analysis of algorithmic fairness, privacy, surveillance, autonomy, and transparency. Compared with the EU AI Act, OECD AI Principles, and UNESCO Recommendation on AI Ethics.

Jobs and Skills	Net employment effects, skills requirements, reskilling imperatives, and social protection needs are assessed for each technology and sector.
Data Sovereignty	Data governance, localization, cross-border flows, and the risks of data colonialism are examined. The Malabo Convention, AU Data Policy Framework, GDPR influences, and national data protection laws are analyzed.
Indigenous Knowledge	The integration of indigenous and local knowledge with emerging technologies—from agricultural practices to traditional medicine to ecological management—is highlighted as both an ethical imperative and a source of innovation.

Source: Author Compilation, based on GSMA, 2023; World Bank, 2024; IRENA; AU, 2020; ILO; UNESCO, 2021; ITU, 2023; Research ICT Africa, 2023; UNCTAD, 2021; OECD, 2023; WIPO, 2023; Ostrom, 2005; WEF, 2020; AfDB, 2023

1.3 Definition of Emerging, Disruptive, and Frontier Technologies

Clear terminology is crucial for effective discussions among diverse stakeholders—policymakers, researchers, entrepreneurs, civil society, development practitioners, and citizens. The terms "emerging," "disruptive," "frontier," and "Industry 4.0" technologies are frequently used interchangeably in policy discussions, but as shown in Table 1.3, they refer to distinct yet overlapping phenomena with different implications for strategy, investment, and governance.

In the United States, the term "emerging and disruptive technologies (EDTs)" is used in its technology strategy, innovation policy, and defense planning. Here, the term EDT refers to new or rapidly advancing technologies that have the potential to significantly transform economic activity, national security, military capabilities, and societal systems.

Table 1.3. Classification of Technology Categories

Category	Definition	Key Characteristics & African Examples
Emerging Technologies	Innovations relatively underdeveloped but with potential to alter the socio-economic landscape within 5–15 years (Rotolo et al., 2015).	Radical novelty; fast growth; coherent development across domains; prominent potential impact; high uncertainty. Examples: AI-powered diagnostics, agricultural drones, blockchain for land registries.
Disruptive Technologies	Innovations that significantly alter or displace existing technologies, products, or systems, often starting in underserved markets (Christensen, 1997).	Initially underperform on mainstream metrics; serve overlooked segments; improve over time; eventually displace incumbents. Classic African example: mobile money (M-PESA, MTN MoMo, Wave, Orange Money, Telebirr) disrupting formal banking.
Frontier Technologies	Cutting-edge innovations at the boundary of current knowledge and capabilities (UNCTAD, 2021).	Defined by position at the forefront of innovation rather than market impact. Include AI, quantum computing, synthetic biology, and advanced materials. Represents both the leapfrogging opportunity and the capability challenge for Africa.
Industry 4.0 Technologies	The suite of digital, physical, and biological innovations driving the Fourth Industrial Revolution (Schwab, 2016).	Distinguished by convergence: cyber-physical systems, biologically enhanced digital tools, unprecedented connectivity. Less about replicating advanced-country automation and more about leveraging convergence for contextual solutions.

Sources: Adapted from Rotolo et al. (2015); Christensen (1997); Schwab (2016); UNCTAD (2021).

This book uses the term "emerging technologies" to refer to artificial intelligence, additive manufacturing, autonomous vehicles and drones, blockchain, big data analytics, IoT, robotics, augmented and virtual reality, as well as distributed renewable energy systems, all viewed through the lens of sustainable development in Africa.

These core technologies are interconnected rather than used separately, working together seamlessly. For example, "AI for agriculture" typically involves satellite or IoT sensing, mobile connectivity, data platforms, and payment systems to turn analytics into actionable decisions for farmers.

The main purpose of writing this volume is to promote the interlinkages among these emerging technologies. African countries need to keep an eye on these key technologies and develop capabilities to achieve broader economic and social benefits. While focusing on AI is important, greater benefits are possible only when countries build expertise across a range of emerging technologies, including blockchain, biotechnology, AI, autonomous systems, IoT, and advanced energy solutions.

The book also explores the critical infrastructure needed to implement these technologies: broadband connectivity, data centers and cloud computing, energy systems, and soft skills such as digital literacy, research capacity, governance and regulatory frameworks, and financing mechanisms.

1.4 Caution on the Limitations

Analytical work on emerging technologies in Africa faces a series of challenges, ranging from data limitations to a rapidly evolving technological landscape and the continent's diversity.

Data Availability and Quality Constraints

Data availability and quality vary widely across African countries and socio-economic sectors in terms of technology adoption. Although improvements in data on ICT/digital infrastructure have been achieved through efforts by the International Telecommunication Union (ITU), GSMA, Research ICT Africa, and the World Bank, detailed subnational data—such as urban–rural, gender-disaggregated, and income-disaggregated figures—are still scarce. Data on innovation and R&D are especially limited, as many African countries do not regularly conduct R&D surveys. Publications and databases often miss informal or locally specific innovations.

Data on technology adoption beyond basic connectivity indicators—like mobile subscriptions and Internet penetration—are generally unavailable or unreliable. Tracking financial flows, such as venture capital, remittances, and informal investments, is difficult. Gender-disaggregated technology data are improving through the work of organizations such as GSMA and Research ICT Africa. However, in most countries, they still fall short of enabling comprehensive gender analysis.

These data gaps have practical implications for the analysis in this volume. Where data are strong (e.g., mobile money adoption, broadband coverage, venture capital flows), the analysis can be precise. Where data are weak (e.g., AI adoption, blockchain deployments, IoT research), the analysis depends more on case studies, expert opinions, and qualitative evidence. Throughout the book, data sources are cited, and confidence levels are indicated to help readers gauge their reliance on specific findings.

Rapidly Evolving Technological Landscape

Emerging technologies are inherently fast-changing. By the time a book is published, some technologies will have progressed, new applications will have emerged, some startups will have failed while others will have scaled, and regulatory landscapes will have shifted. This is an inherent limitation of any static publication on rapidly evolving topics.

The book addresses this by emphasizing analytical frameworks and principles that remain relevant even as specific technologies change; by focusing on trajectories and trends rather than snapshots; by offering scenario-based analysis that considers multiple futures; and by identifying indicators and information sources that readers can use to update their analysis. The European Commission handles this challenge through annual or biannual updates to its Digital Economy and Society Index (DESI). The ITU provides annual data updates. Readers of this volume are encouraged to supplement it with these regularly updated sources.

Heterogeneity Across African Contexts

Africa comprises 55 African Union members, over 2,000 languages, a range of climate zones from the Sahara Desert to rainforests near the equator, populations ranging from less than 1 million (Seychelles, Cabo Verde) to over 200 million (Nigeria), and income levels from the least developed to upper-middle-income. Any continental-level analysis necessarily simplifies this diversity.

The book addresses this by providing country-specific examples and case studies from all regions, using regional disaggregation where data permit; by flagging when continental averages mask significant variation; and by emphasizing that the frameworks and principles presented need to be adapted to specific country or subnational contexts.

This heterogeneity is itself an asset: the diversity of African contexts creates natural experiments that facilitate learning. Rwanda's centralized digital transformation strategy contrasts with Nigeria's market-driven, decentralized approach. Kenya's fintech regulatory innovation offers lessons for Ghana's drone regulation and South Africa's renewable energy procurement. These contrasts enable comparative analysis that enriches the volume's findings.

1.5. Conclusion

This chapter introduces the reader to the purpose, analytical framework, temporal scope, and data limitations that shape its analysis. Chapter 2 discusses why emerging technologies are critical for Africa. The subsequent chapters examine the development impact of specific emerging technologies, their alignment with AU Agenda 2063, the enabling conditions and challenges, and strategic pathways forward.

References

African Development Bank. (2023). African Economic Outlook 2023: Mobilizing private sector financing for climate and green growth in Africa. African Development Bank Group.

African Union. (2020). The digital transformation strategy for Africa (2020–2030). African Union.

Asian Development Bank. (2022). Asian Development Outlook 2022: Mobilizing taxes for development. Asian Development Bank.

Christensen, C. M. (1997). The innovator's dilemma: When new technologies cause great firms to fail. Harvard Business School Press.

Economic Commission for Latin America and the Caribbean. (2022). Digital Technologies for a New Future. ECLAC.

European Commission. (2021). 2030 Digital Compass: The European way for the Digital Decade. European Commission.

European Commission. (2023). Digital Economy and Society Index (DESI) 2023. European Commission.

Geels, F. W. (2004). From sectoral systems of innovation to socio-technical systems. Research Policy, 33(6–7), 897–920.

GSMA. (2023). State of the Industry Report on Mobile Money 2023. GSMA.

GSMA. (2023). The Mobile Gender Gap Report 2023. GSMA Connected Women.

International Telecommunication Union. (2023). Measuring Digital Development: Facts and Figures 2023. ITU Publications.

Lundvall, B.-Å. (Ed.). (2010). National systems of innovation: Toward a theory of innovation and interactive learning. Anthem Press.

Nussbaum, M. C. (2011). Creating Capabilities: The Human Development Approach. Harvard University Press.

Organization for Economic Co-operation and Development. (2023). OECD Science, Technology and Innovation Outlook 2023. OECD Publishing.

Ostrom, E. (2005). Understanding institutional diversity. Princeton University Press.

Research ICT Africa. (2023). After Access Survey: ICT access and use in Africa. Research ICT Africa.

Rogers, E. M. (2003). Diffusion of innovations (5th ed.). Free Press.

Schwab, K. (2016). The fourth industrial revolution. World Economic Forum.

Sen, A. (1999). Development as freedom. Oxford University Press.

Smart Africa. (2023). Smart Africa Alliance: Progress report 2023. Smart Africa Secretariat.

UNESCO. (2021). UNESCO science report: The race against time for smarter development. UNESCO Publishing.

United Nations Conference on Trade and Development. (2021). Technology and innovation report 2021. United Nations Publications.

United Nations Department of Economic and Social Affairs. (2023). World population prospects 2023. United Nations.

World Bank. (2024). World development indicators 2024. World Bank Publications.

World Economic Forum. (2020). The future of the global financial system. World Economic Forum.

World Intellectual Property Organization. (2023). World Intellectual Property Report 2023. WIPO Publication.

Chapter 2: Emerging Technologies Imperatives, Transitions and Risks for Africa

2.1 Introduction

This chapter lays out the conceptual and contextual basis for understanding why emerging technologies are crucial for Africa and how they relate to the continent's development trajectory. It frames them within the global innovation ecosystem and historical technological shifts, and examines Africa's unique position in the global technology landscape. It also introduces the key issues of risk, ethics, and inclusion that will be discussed in more detail in later chapters.

2.2. Emerging Technologies Imperatives for Africa

Africa's engagement with emerging technologies is essential, not optional. The continent's 1.4 billion people, expected to grow to 2.5 billion by 2050 with over 60 percent under age 25, represent both the world's largest future labor force and the most important test for technology-driven inclusive development (UNDESA, 2023). This demographic advantage aligns with the Fourth Industrial Revolution—marked by the merging of digital, biological, and physical technologies—that is transforming economies, labor markets, and social structures (Schwab, 2016).

At the same time, Africa faces persistent structural challenges: gaps in energy, connectivity, and transportation infrastructure, along with limited industrialization, with manufacturing making up only about 10 percent of Gross Domestic Product (GDP), compared to 25 percent in East Asia. Climate vulnerability threatens agricultural livelihoods, and Africans have unequal access to education, healthcare, water, and sanitation (AfDB, 2023; World Bank, 2024). Similar challenges affect parts of South and Southeast Asia, but the scale and simultaneity of Africa's needs are unique.

Figure 2.1. Africa's Population Growth and Youth Share

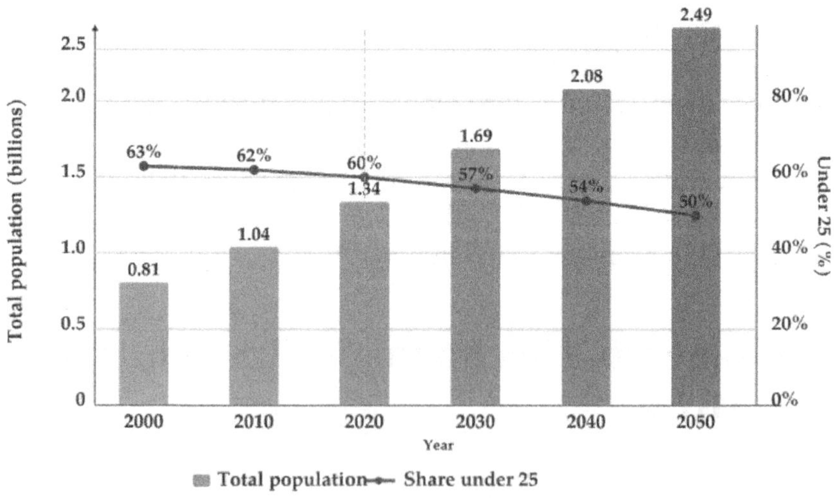

Sources: UN DESA (2023)

Africa's imperative to strategically adopt emerging technologies stems from four interconnected factors shaping its development and future.

Economic Transformation and Structural Change

Emerging technologies, if properly adopted, could transform African economies. The economies in the region remain structurally unbalanced, heavily reliant on primary commodity exports. Manufacturing's share of GDP has stalled or decreased in most African countries over the past 20 years—a pattern known as "premature deindustrialization" that sharply contrasts with the manufacturing-led growth seen in East Asian economies such as the Republic of Korea, Vietnam, and China (Rodrik, 2016; AfDB, 2023). Latin America and the Caribbean face a similar issue: the region's manufacturing sector has also shrunk as a portion of GDP, dropping from around 17 percent in 2000 to 13 percent in 2022 (ECLAC, 2022). This global trend risks leaving 10–12 million young Africans entering the labor market each year without enough productive jobs.

Figure 2.2. Regional Manufacturing Share of GDP

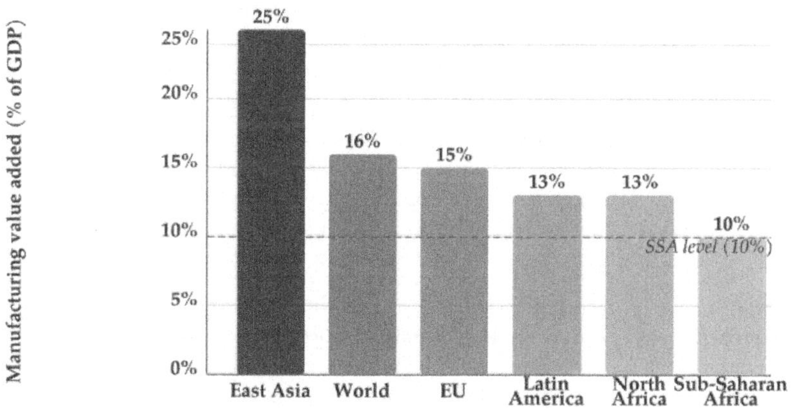

Sources: World Bank (2024); AfDB (2023)

Emerging technologies provide multiple ways to promote economic change. From a technical perspective, digital platforms help enterprises access global markets; robots can improve manufacturing competitiveness. AI can greatly speed up responses to climate change, improve food security, enhance educational outcomes, and advance healthcare delivery, while blockchain can improve supply chain transparency.

Table 2.1. Technology Domains and Potential Economic Transformation Pathways in Africa

Technology Domain	Transformation Pathway and African Examples
Digital Platforms	Digital platforms enable entrepreneurs and SMEs to access global markets, reducing barriers of distance and intermediation. Examples in Africa include Jumia (pan-African e-commerce), Flutterwave and Paystack (payments), and Kenya's business process outsourcing sector. Brazil's Nubank and India's Jio Platforms illustrate comparable platform-led transformation in other emerging markets (Manyika et al., 2016).
Automation & Robotics	Automation enhances manufacturing competitiveness by addressing quality, productivity, and global value-chain integration. South Africa's automotive sector and Egypt's textile industry are early adopters. While automation raises concerns about job displacement, it can also attract investment and create adjacent employment in design, programming, and maintenance (Hallward-Driemeier & Nayyar, 2017).
AI & Big Data	AI and big data optimize resource allocation from agriculture to healthcare. Rwanda's use of AI-driven health analytics and Nigeria's credit-scoring platforms demonstrate early applications. China's agricultural AI deployments in precision farming offer models for technology transfer to African contexts (Vinuesa et al., 2020).
Additive Manufacturing	Additive manufacturing (3D printing) enables the local production of spare parts, medical devices, and customized products, reducing import dependence. Kenya's Gearbox initiative and South Africa's 3D-printed medical implants are pioneering cases (Campbell et al., 2018).
Blockchain & DLT	Digital Ledger Technologies offers supply-chain transparency, enabling African producers to demonstrate provenance and quality. Ghana's cocoa traceability initiatives and Ethiopian coffee supply-chain pilots illustrate development-relevant applications (Kamble et al., 2020).

Sources: Author compilation based on AfDB (2023); World Bank (2024); UNCTAD (2021).

From a job standpoint, emerging technologies can facilitate a shift in employment toward higher-income, more productive service industries—such as digital services exports, the platform economy, or technology-driven manufacturing.

Table 2.2. Structural Transformation Pathways Enabled by Emerging Technologies

Pathway	Description and Requirements	African and Global Examples
Digital Services Export	Export software, BPO, digital content, and professional services. Requires strong education, reliable connectivity, and competitive labour costs.	Rwanda, Kenya, and South Africa are comparable to India's IT services model and the Philippines' BPO industry.
Tech-Enhanced Manufacturing	Smart manufacturing, 3D printing, robotics for precision tasks. Enables competition in specialised or customised products.	South Africa automotive, Egypt textiles, Ethiopia industrial parks; comparable to Viet Nam's electronics assembly integration.
Agricultural Modernisation	Precision agriculture, processing, and digital market platforms. Requires cold chains, skills, and market access.	Twiga Foods (Kenya), Agricool (Nigeria), Apollo Agriculture are comparable to Brazil's agri-tech ecosystem.
Platform Economy	Ride-hailing, freelancing, e-commerce, and digital content. Generates income but raises concerns about precarity.	SafeBoda (Uganda), M-Pesa ecosystem, Andela; comparable to Indonesia's Gojek and Brazil's iFood.
Resource-Based Industrialization	Local processing of minerals, energy, and agricultural feedstocks with advanced technologies.	DRC cobalt processing ambitions, Morocco phosphate value addition, Namibia green hydrogen; comparable to Chile's lithium strategy.

Sources: Author compilation based on Rodrik (2016); McMillan & Rodrik (2011); AfDB (2023).

These pathways are not mutually exclusive; successful strategies likely combine elements tailored to different country contexts. Furthermore, transformation should be evaluated not only by GDP growth but also by employment creation assessment (especially for youth and women), income distribution, environmental sustainability, and resilience to shocks.

Social Development and Service Delivery

Emerging technologies have significant potential to improve health, education, and public service delivery. About 600 million Africans lack access to electricity, 400 million lack reliable clean water, and health systems are severely under-resourced, with an average of 2 physicians per 10,000 people compared to 32 in Europe and 18 worldwide (WHO, 2021; IEA, 2022). Education systems face issues related to quality, relevance, and access, particularly in rural areas. These gaps are not unique to Africa—parts of South Asia and the Pacific Islands experience similar shortages—but their scale across the continent is unparalleled.

Emerging technologies offer scalable solutions to these persistent challenges:

- **Telemedicine and AI-driven diagnostics** have expanded access to healthcare in underserved rural areas. Zipline's drone delivery of blood and vaccines in Rwanda and Ghana, Babyl Health's AI triage in Rwanda, and mPharma's pharmacy management platform across West Africa demonstrate what is possible. India's e-Sanjeevani telemedicine platform, which facilitated over 100 million consultations by 2023, serves as a model for scaling such services (Amukele et al., 2017).

- **Internet of Things (IoT) sensors** enable real-time monitoring of water quality, distribution systems, and agricultural conditions. In Senegal, IoT-based water monitoring has reduced system losses by up to 30 percent in pilot programs. Similar implementations in Bangladesh and the Philippines show cross-regional relevance of IoT adoption (Adu-Manu et al., 2017).

- **Solar mini-grids and pay-as-you-go (PAYG) renewable energy systems** have connected millions of households and businesses that were previously off-grid across East and West Africa. Companies such as M-KOPA (Kenya), BBOXX (Pan-African), d.light (global), and Zola Electric (Tanzania and Nigeria) have demonstrated how distributed technologies can address infrastructure gaps more quickly than grid extension (Blodgett et al., 2017).

- **Digital learning platforms** expand access to quality education, especially in STEM fields. Eneza Education (Kenya), Ubongo (Tanzania), and the Andela distributed talent model showcase different approaches to traditional education. India's DIKSHA platform provides comparative frameworks for nationwide digital education strategies (Bates, 2015).

Figure 2.3. Infrastructure Deficit in Africa in the 2020s

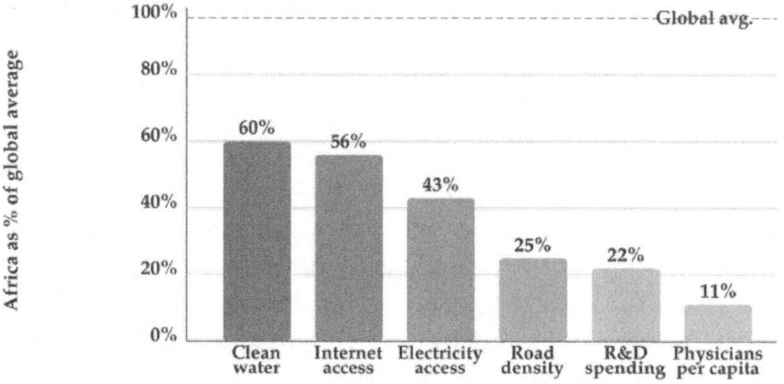

Sources: ITU (2023); IEA (2022); WHO (2021); UNESCO (2021)

Climate Resilience and Environmental Sustainability

Climate adaptation is another area that could benefit from increased use of emerging technologies. Africa accounts for around 4 percent of global greenhouse gas emissions. Yet, it faces disproportionate climate impacts: more frequent and severe droughts and floods, changing rainfall patterns that disrupt agricultural schedules, rising temperatures that threaten crop yields, and coastal erosion that jeopardizes infrastructure (Intergovernmental Panel on Climate Change, 2022).

Climate change exacerbates existing vulnerabilities in food security, water access, and health outcomes. Small Island Developing States across Africa, the Caribbean, and the Pacific face similar threats from rising sea levels, emphasizing the global justice aspect of the climate–technology dynamic. Some ways emerging technologies can be used for climate adaptation include:

- **Climate modeling and big data analytics** improve weather forecasting, early warning systems, and climate-smart agricultural planning. Morocco's national meteorological service and the Inter-Governmental Authority for Development (IGAD) Climate Prediction and Applications Centre (ICPAC) for East Africa are some examples of climate modeling (Thornton et al., 2014).

- **Precision agriculture**—using soil sensors, satellite imagery, and variable-rate application systems—enhances input efficiency and improves yields. Apollo Agriculture (Kenya), Zenvus (Nigeria), and AeroFarms in Sub-Saharan Africa are implementing these technologies for smallholder farmers (Gebbers & Adamchuk, 2010).

- **Renewable energy**—Africa's renewable potential is immense—10 terawatts of solar capacity alone. Kenya's Lake Turkana Wind Power project (310 MW), Morocco's Noor-Ouarzazate solar complex (580 MW), and South Africa's Renewable Energy Independent Power Producer Procurement Programme (REIPPPP) showcase utility-scale ambitions at national levels (International Renewable Energy Agency, 2022).

- **Environmental monitoring** includes satellite remote sensing, IoT sensor networks, and drones that support tracking deforestation, wildlife populations, and pollution. The Regional Centre for Mapping of Resources for Development (RCMRD) in Kenya and Digital Earth Africa deliver continental-scale environmental monitoring (Turner et al., 2015).

Global Competitiveness and Technological Sovereignty

Africa has an equal opportunity to improve its technological competitiveness by replacing outdated applications with open standards. Countries that build local technology capabilities—such as research and development, a skilled workforce, innovation systems, and supporting infrastructure—gain a larger share of value creation. The European Union's pursuit of "digital sovereignty" through its Digital Decade strategy and significant investments in semiconductor manufacturing exemplify this approach. India's Digital India program and China's Made in China 2025 strategy show how national technology policies can drive structural transformation (Baldwin, 2016).

Africa's strategic technology engagement is essential across several dimensions:

- **Positioning in global value chains:** As automation decreases labor-cost advantages, African countries must prioritize innovation, quality, and speed to market. Ethiopia's digital-economy plan for 2025–2030 and Rwanda's initiatives as an ICT hub demonstrate national strategies that position these countries. The African Continental Free Trade Area (AfCFTA) creates a market of 1.4 billion people that, if supported by digital trade infrastructure, could attract technology investments comparable to ASEAN's digital economy integration (Baldwin, 2016). Africa's startup ecosystem is small but growing, already attracting nearly US$3 billion in annual investment.

- **Data sovereignty and governance:** Data is increasingly regarded as a strategic asset. The African Union's Convention on Cyber Security and Personal Data Protection (Malabo Convention) and national laws in Kenya, South Africa, Nigeria, and Ghana are establishing frameworks for data sovereignty and governance. The AU Data Policy Framework guides a governance framework that balances localization with data flows. The Framework recognizes that excessive data localization can hamper digital trade, innovation, and regional integration.

- **Building on comparative advantages:** Africa possesses unique assets—abundant renewable energy, critical minerals needed for tech hardware (cobalt from the Democratic Republic of the Congo, lithium reserves in Zimbabwe and Mali, rare earth elements in Namibia and South Africa), linguistic and cultural diversity valuable for AI training, and a large young population that is adaptable to digital technologies (Signé & Johnson, 2018).

- **South–South cooperation and knowledge sharing:** Africa's development challenges are common with other regions in the Global South. India's Stack-based Digital Public infrastructure (Aadhaar identity, UPI payments) and Brazil's Pix instant-payment system serve as models —and also offer cautionary examples—for African digital public infrastructure strategies (Quadir, 2020).

Figure 2.4. The Growing Startup Ecosystem in Africa. Venture Capital Investment in African Technology Startups (2017–2024)

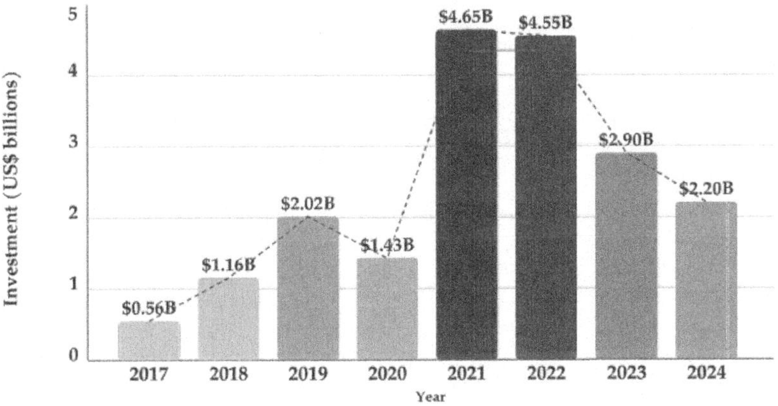

Sources: Partech Africa, Briter Bridges, Africa: The Big Deal estimates

These four imperatives—economic transformation, social development, climate resilience, and global competitiveness—highlight why Africa cannot afford to be passive in its engagement with emerging technologies. The way forward requires understanding not only the technologies themselves but also their interconnectedness, enabling conditions, and the governance structures needed to manage risks while optimizing developmental benefits.

2.3 Technology Lifecycles and Adoption Curves

Technologies pass through predictable lifecycle stages, each marked by specific features and impacts on adopters (Utterback & Abernathy, 1975). The S-curve model illustrates adoption patterns: slow initial uptake as early users experiment with the technology, rapid growth as the broader market adopts it, and eventually saturation (Rogers, 2003). However, adoption rates vary greatly between countries and populations, depending on income, education, infrastructure, institutions, and cultural factors.

Figure 2.5. Technology Lifecycle S-Curve with African and Global Examples

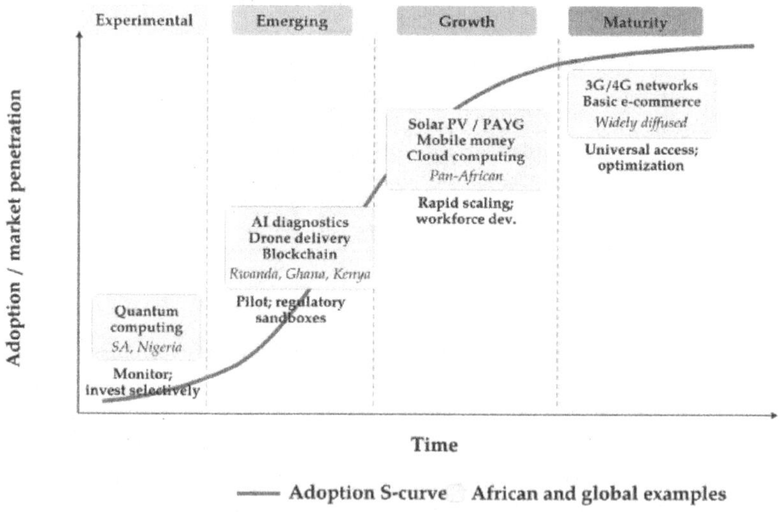

Source: Adapted from Rogers (2003) and Utterback & Abernathy (1975)

In the African context, effective technology strategies should include concurrent efforts across various stages: adopting established technologies for immediate benefits (such as 4G and 5G connectivity, solar home systems), investing in emerging technologies with clear development significance (like AI for healthcare, IoT for farming, blockchain for land records), and keeping an eye on experimental technologies that might become essential in the future (green hydrogen, quantum computing, synthetic biology).

Table 2.3. Technology Lifecycle Stages: Features, African Examples, and Strategic Implications

Stage	Characteristics	African Examples	Strategy Implication
Experimental	R&D dominant; high cost; extreme uncertainty; limited applications.	Quantum computing research (South Africa, Nigeria); general AI adoption for selected languages.	Monitor; invest selectively in basic research and international collaboration.
Emerging	Technical viability demonstrated; early adopters experimenting; standards forming; costs declining.	AI diagnostics; drone delivery (Zipline); blockchain use in supply chains.	Strategic early investment; pilot programmes; regulatory sandboxes.
Growth	Adoption accelerating; ecosystems developing; competition intensifying; dominant designs emerging.	Solar PV, mobile money, cloud computing.	Rapid scaling; infrastructure investment; workforce development; competition policy.
Maturity	Widely diffused; incremental improvements; low risk but limited competitive advantage.	3G/4G networks; basic e-commerce; conventional solar PV.	Universal access; optimization; transition planning to next generation like 5G.

Sources: Adapted from Utterback & Abernathy (1975); Rogers (2003); UNCTAD (2021)

This approach aligns with the European Commission's "Key Enabling Technologies" strategy and Singapore's Research, Innovation, and Enterprise plans, which combine short-term applications with long-term capability development.

2.4 Technology Transitions and Africa's Context

Each major stage of economic growth has been driven by technological change: the agricultural revolution enabled settled civilizations; the first industrial revolution transformed manufacturing with steam power; the second introduced electricity and mass production; the third brought computing and telecommunications. Now, the Fourth Industrial Revolution promises—or threatens—another fundamental shift through the integration of digital, physical, and biological technologies (Schwab, 2016).

The history of industrialization shows both the transformative power of technology and the challenges that come with technology-driven growth. Britain's Industrial Revolution relied on coal, steam, and mechanized textiles. Germany focused on chemicals, electrical equipment, and scientific innovation. The United States used its abundant resources, mass production techniques, and a large domestic market. Japan combined imported technology with local adaptation and strong government coordination (Landes, 1969; Allen, 2009). These examples provide several lessons relevant to the African context today.

- **There is no one-size-fits-all approach**—countries have industrialized using different strategies tailored to their specific circumstances, resources, and historical contexts. Africa's development path should not copy any previous model in its entirety.

- **Technology adoption requires complementary capabilities**, including infrastructure, human capital, and institutions. China's rapid industrialization, for example, was preceded by decades of investment in basic education and rural infrastructure.

- **Late industrializers face unique challenges**—countries can import and adapt existing technologies, but they also compete with established powers and often face less favorable terms of technology transfer.

- **Structural transformation requires more than technology**—shifting from low-productivity to high-productivity activities requires simultaneous investments in infrastructure, skills, institutions, and supportive policies (Rodrik, 2013).

The East Asian developmental states—the Republic of Korea, Taiwan, and Singapore—demonstrated that rapid industrialization can be achieved through strategic technology acquisition, substantial investment in education, and export-focused manufacturing. However, replicating this model in Africa faces significant challenges because automation can erode labor-cost advantages, many African countries lack comparable state capacity, resource endowments differ, and Industry 4.0 requires capabilities that differ from those used by East Asian countries in traditional manufacturing (Amsden, 1989; Wade, 1990).

Africa is a diverse continent comprising 55 African Union Member States, each with significant differences in income, institutions, infrastructure, and technological readiness. This diversity has important implications for how emerging technologies are adopted and used.

Table 2.4. Africa's Socio-Economic Baseline and Implications for Technology Adoption

Dimension	Key Facts	Implications for Technology
Economic Structure	GDP ~$3 trillion (3% of global GDP for 17% of population). GDP per capita ranges from <$500 (Burundi, South Sudan) to >$10,000 (Mauritius, Seychelles)—heavy commodity dependence.	Limited industrial base, but fewer legacy systems to replace. Small national markets, but AfCFTA creates a combined market of 1.4 billion.
Demographics	Median age 19 (vs. 38 in Europe). Life expectancy is 64 years. Rapid urbanisation. Expanded enrolment but poor learning outcomes.	The young population is adaptable to digital technology but requires significant investment in skills. Urbanisation creates density for tech deployment.
Infrastructure	~600 million without electricity; 37% internet access (vs. 66% globally); low road density; inadequate port/airport infrastructure.	Deficits constrain deployment but create opportunities for distributed, leapfrogging solutions (e.g., off-grid solar, satellite broadband).
Governance	Wide variation: Botswana, Mauritius, Rwanda, and Cape Verde score well; others face weak institutions, corruption, and instability.	Weak institutions hinder governance but create space for regulatory experimentation (sandboxes). Trust is fundamental to adoption.
Innovation Capacity	R&D ~0.4% of GDP (vs. global 1.8%). Innovation hubs are proliferating in Kenya, Nigeria, South Africa, Egypt, and Ghana. Mobile money is a global model.	Low formal R&D but vibrant startup ecosystems. Brain drain is a challenge; diaspora is an asset for knowledge transfer.

Sources: AfDB (2023); World Bank (2024); ITU (2023); UNESCO (2021).

2.5 Positioning Africa in the Global Emerging Technology Innovation Ecosystem

Global innovation systems are largely geographically concentrated. The United States, the EU, China, Japan, and the Republic of Korea dominate research spending, patent filings, and the number of tech companies. Africa's role so far is mainly peripheral—mostly as a consumer of imported technologies and a supplier of raw materials. Still, innovation strengths are beginning to emerge in specific areas where African innovators tackle urgent local needs.

- **Mobile money and fintech:** African innovators have developed solutions now adopted worldwide, with mobile money apps used in many countries. Nigeria's fintech scene attracted over $1.3 billion in venture capital in 2022 alone.

- **Drone delivery and aerial logistics:** Rwanda and Ghana are at the forefront of medical drone delivery worldwide. Ghana's regulatory framework, developed with the World Economic Forum, is being reviewed by regulators in Asia and Latin America.

- **Off-grid renewable energy:** African entrepreneurs pioneered PAYG solar models, now replicated in South and Southeast Asia. East Africa's M-KOPA has reached over 3 million homes.

- **Agricultural technology:** Platforms that deliver weather, market, and agronomic details to smallholders via SMS showcase innovation that adapts to infrastructure limitations. Hello Tractor (Nigeria), Twiga Foods

(Kenya), and WeFarm (pan-African) exemplify this niche.

At the same time, Africa's engagement with technology takes place within a polycrisis—multiple interconnected global crises that can accelerate negative impacts (Lawrence et al., 2022). The climate crisis, health vulnerabilities, and ongoing inequality and exclusion shape the immediate environment. This implies that technologies that worsen these crises—through high energy use, environmental damage, or concentrated benefits—are unacceptable, regardless of their efficiency.

Emerging technologies that address multiple crises simultaneously—such as renewable energy for climate mitigation and energy access, telemedicine for health and rural inclusion, and digital platforms for jobs and economic opportunities—should therefore be prioritized and expanded. This approach aligns with the EU's "twin transition" framework, which combines the green and digital agendas, and with the Caribbean's focus on climate-resilient digital infrastructure.

2.6 Leapfrogging, Catch-Up, and Structural Transformation

Discussions about Africa catching up with technological development, bypassing intermediate stages, and leapfrogging directly to advanced technologies have taken place over the past twenty years. However, evidence shows that leapfrogging is possible only in specific areas under certain conditions and requires intentional strategies, additional investments, and realistic assessments of prerequisites. Leapfrogging involves adopting advanced technologies without progressing through the intermediate stages that earlier adopters experienced (Steinmueller, 2001).

Leapfrogging can be categorized into technological leapfrogging (adopting new technologies without relying on older ones), stage-skipping leapfrogging (skipping development stages), and path-creating leapfrogging (developing innovative solutions that may serve as models for others). The opposing view emphasizes sequential capability-building through learning-by-doing, exemplified by the Republic of Korea's transition from simple manufactured exports to semiconductors and shipbuilding over three decades (Amsden, 1989; Lee, 2013). The best approach lies somewhere between these perspectives; success depends on the characteristics of the technology, market structures, institutional frameworks, and development objectives.

Mobile money is Africa's top success story of leapfrogging. Kenya's M-PESA, launched in 2007, enabled millions of previously unbanked Kenyans to access financial services through their mobile phones; this model has been replicated elsewhere in Africa. By 2023, there were over 500 million mobile money accounts across Sub-Saharan Africa (GSMA, 2023). Other African leapfrogging successes include solar home systems and mini-grids (M-KOPA, BBOXX), agricultural information platforms (Esoko, WeFarm), and digital identity systems (Ethiopia, Ghana, Kenya, and Nigeria).

India's Unified Payments Interface (UPI), which handled more than 10 billion transactions per month by 2023, reflects a similar but structurally different leapfrogging success — based on national digital public infrastructure rather than mobile operator-led innovation. Brazil's Pix system, which reached 150 million users within two years of its launch, presents yet another model.

These successes share common features: they address urgent needs, leverage widespread mobile networks, use simple, affordable interfaces, are supported by favorable regulations and strong championship, and rely on financially viable business models. High mobile phone penetration, significant unmet demand, supportive regulation, a broad agent network, and user-focused design drove the progress of mobile money in Africa.

Figure 2.6. Prerequisites for Successful Technology Leapfrogging

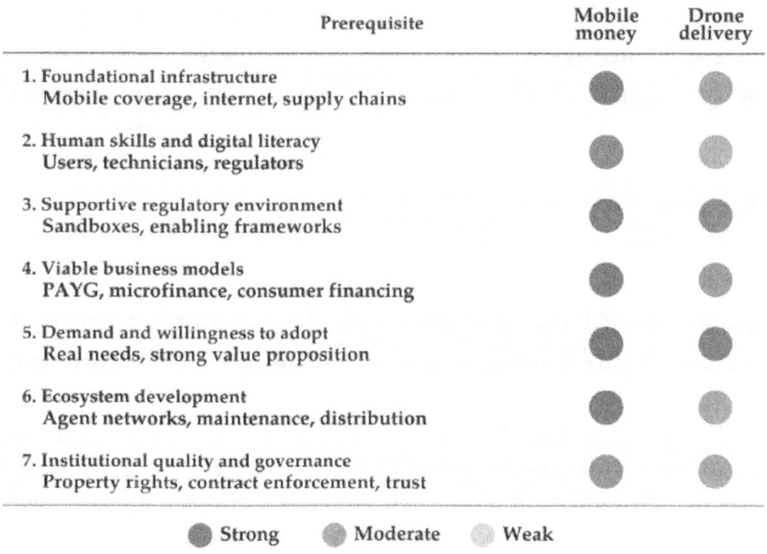

Source: Author Analysis

Analysis of successful and failed leapfrogging attempts across Africa, Asia, and Latin America reveals seven critical prerequisites:

- **Foundational infrastructure and connectivity are essential.** Even leapfrogging technologies need basic infrastructure—mobile coverage for fintech, Internet access for telemedicine, and supply chains for solar systems.

- **Essential human skills.** Users require digital literacy; providers need technical expertise; regulators must understand the technologies.

- **Creating a supportive regulatory environment.** Regulatory sandboxes—successfully tested in Kenya (fintech), Rwanda (drones), and the United Kingdom (financial services)—enable controlled experimentation.

- **Viable business models and financing options.** Pay-as-you-go (PAYG) models, microfinance, and consumer financing have been essential to the program's success. The EU's InvestEU program and India's Startup India provide comparable financing frameworks.

- **Demand and willingness to adopt.** Technologies succeed when they meet real needs with strong value offers. User-centered design and local context are crucial.

- **Ecosystem development.** Technologies need spare parts, maintenance, distribution networks, and complementary services. Morocco's Technopolies of Casablanca, Agadir, Rabat, and Tetouan; Tunisia's El Gazala and Novation City in Sousse; Egypt's Maadi, Smart City, and Knowledge City; Kenya's "Silicon Savannah"; and Nigeria's Yaba Tech Cluster demonstrate the impact of ecosystems on tech development..

- **Institutional quality and governance**. Property rights, contract enforcement, and low corruption foster environments conducive to investment and trust—the foundation of technology adoption.

2.7 Risks, Ethics, and Inclusion in Technology Adoption

While emerging technologies offer advantages, they also pose risks that are widely discussed throughout the book, with special treatment in Chapter 9. Enthusiasm for technology as a development solution must therefore be balanced with careful analysis of risks, ethical concerns, and distributional impacts.

Table 2.5. The Dual Nature of Technology: Enabler and Disruptor

Domain	Opportunity	Risk
Automation & Employment	Increases productivity; creates new industries and occupations.	Displaces workers in automatable tasks. Acute for Africa, where 10–12 million youth enter labour markets annually.
Digital Platforms	Reduce transaction costs; enable market participation by small businesses, rural producers, informal workers.	Concentrate market power, extract value through intermediation, and exploit data asymmetries.
Surveillance & Data	Enable disease surveillance, targeted social programmes, and evidence-based policymaking.	Enable mass surveillance, privacy violations, social control. Especially dangerous with weak institutions.
Algorithmic Decisions	Process vast data; identify patterns humans miss; improve decision quality.	Perpetuate biases; produce opaque decisions. Facial recognition has higher error rates for darker skin tones; credit algorithms may encode discrimination.
Environmental Impact	Dematerialisation; resource-use optimisation; renewable energy integration.	Energy consumption (data centres); e-waste from short device lifespans; critical mineral extraction with environmental and social costs.

Sources: Author compilation based on Vinuesa et al. (2020); UNCTAD (2021); Hallward-Driemeier & Nayyar (2017).

General-purpose emerging technologies like AI pose significant ethical challenges, including data privacy, algorithmic transparency, and ethical decision-making. Applications such as biometrics and AI collect sensitive personal or genetic data, raising concerns about consent and excessive surveillance, especially in the absence of strong regulations. AI systems trained on non-African datasets can generate biased results, leading to unfair healthcare, financial, and employment outcomes that deepen marginalization.

Addressing technological risks requires grounding ethics in African values, contexts, and priorities rather than uncritically adopting frameworks from elsewhere. The EU's AI Act, the OECD AI Principles, UNESCO's Ethical AI framework, and China's New Generation AI Governance Principles serve as reference points. However, African ethical frameworks must incorporate the continent's unique philosophical traditions and development goals.

The key principles include:

- **Human dignity and ubuntu.** African philosophy highlights interconnectedness, community, and shared humanity ("I am because we are"). Technologies should support rather than undermine human dignity and strengthen rather than fracture communities.

- **Justice and equity.** Given Africa's history of exploitation and ongoing inequalities, the deployment of technology must actively support justice. This includes benefit-sharing systems, fair pricing, universal service commitments, and accessible design.

- **Sovereignty and self-determination.** African countries need to retain control over their technology choices, data policies, and development strategies. The African Union's Digital Transformation Strategy, the AU Data Policy Framework and the Malabo Convention represent the continental frameworks for digital sovereignty.

- **Cultural preservation and diversity.** Africa's linguistic diversity, with over 2,000 languages, is a crucial part of identity and knowledge. Technologies should support this diversity by using AI for African languages (as exemplified by Lelapa AI, Masakhane NLP), digitally preserving cultural heritage, and enabling content-creation platforms.

- **Intergenerational responsibility.** With the world's youngest population, African technology choices have significant implications for future generations. Decisions should prioritize long-term sustainability and avoid harmful path dependencies.

- **Transparency and accountability.** Technology systems that impact the public interest must be transparent and accountable. This includes explainable AI in high-stakes decisions, public disclosure of algorithms used in public services, and mechanisms for redress.

Apart from the inherent risks of emerging technologies, significant gaps in infrastructure, such as Internet connectivity and electricity, create unequal access to emerging technologies, thus limiting their benefits for everyone. Inclusive innovation should involve developing and deploying technologies that serve all parts of society, especially those traditionally excluded (Heeks et al., 2014). If emerging technologies only benefit elites while bypassing or harming the poor, rural populations, women, or marginalized groups, they do not foster inclusive development.

Achieving true inclusion requires deliberate actions across multiple areas: access (affordability, connectivity, devices), usability (local languages, simple designs, cultural relevance), relevance (participatory design that addresses real needs), agency (empowering users as innovators rather than passive consumers), equitable benefit sharing (local hiring, SME involvement, progressive taxation), and gender inclusion (addressing affordability issues, social norms, digital skills gaps, and online safety). Inclusive technology strategies should explicitly target these divides through specific interventions, as shown by initiatives such as the EU's Digital Skills and Jobs Coalition and India's Common Service Centres, which provide digital access to over 400,000 villages.

2.8 Conclusion

This chapter indicates that Africa's engagement with emerging technologies is both timely and complex. The opportunities are real and significant — boosting economic growth, providing large-scale social services, addressing climate change, and helping Africa compete globally. The risks are also considerable: job losses, greater inequality, environmental damage, privacy concerns, and dependence on technology. Success depends on moving beyond simple optimism or pessimism to strategic, evidence-based approaches that maximize benefits, manage risks, promote inclusion, foster innovation, and develop Africa's independence while engaging with global systems.

Emerging technologies should not be viewed in isolation from one another, but the key capabilities across all ten emerging technologies discussed in this book should be developed regionally and nationally. The remaining chapters explore, in detail, specific technologies, sector applications, enabling conditions, governance frameworks, and capabilities needed to harness their benefits while minimizing their risks.

References

Adu-Manu, K. S., Tapparello, C., Heinzelman, W., Katsriku, F. A., & Abdulai, J. D. (2017). Water quality monitoring using wireless sensor networks. ACM Transactions on Sensor Networks, 13(1), 1–41.

African Development Bank. (2023). African economic outlook 2023. African Development Bank Group.

Allen, R. C. (2009). The British Industrial Revolution in global perspective. Cambridge University Press.

Amukele, T. K., Sokoll, L. J., Pepper, D., Howard, D. P., & Street, J. (2017). Can unmanned aerial systems be used for the routine transport of laboratory specimens? PLOS ONE, 12(7), e0134020.

Amsden, A. H. (1989). Asia's next giant: South Korea and late industrialization. Oxford University Press.

Baldwin, R. (2016). The Great Convergence: Information Technology and the New Globalization. Harvard University Press.

Bates, A. W. (2015). Teaching in a digital age. Tony Bates Associates Ltd.

Blodgett, C., Dauenhauer, P., Louie, H., & Kickham, L. (2017). Accuracy of energy-use surveys in predicting rural mini-grid user consumption. Energy for Sustainable Development, 41, 88–105.

Campbell, T., Williams, C., Ivanova, O., & Garrett, B. (2018). Could 3D printing change the world? Atlantic Council.

Christensen, C. M. (1997). The innovator's dilemma. Harvard Business School Press.

Couldry, N., & Mejias, U. A. (2019). Data colonialism: Rethinking big data's relation to the contemporary subject. Television & New Media, 20(4), 336–349.

Economic Commission for Latin America and the Caribbean. (2022). Digital technologies for a new future. ECLAC.

Gebbers, R., & Adamchuk, V. I. (2010). Precision agriculture and food security. Science, 327(5967), 828–831.

GSMA. (2023). State of the industry report on mobile money 2023. GSMA.
Hallward-Driemeier, M., & Nayyar, G. (2017). Trouble in the making? World Bank Publications.
Heeks, R., Foster, C., & Nugroho, Y. (2014). New models of inclusive innovation for development. Innovation and Development, 4(2), 175–185.
Intergovernmental Panel on Climate Change. (2022). Climate change 2022: Impacts, adaptation and vulnerability. Cambridge University Press.
International Energy Agency. (2022). Africa Energy Outlook 2022. IEA Publications.
International Renewable Energy Agency. (2022). Renewable capacity statistics 2022. IRENA.
International Telecommunication Union. (2023). Measuring digital development: Facts and figures 2023. ITU.
Kamble, S. S., Gunasekaran, A., & Gawankar, S. A. (2020). Achieving sustainable performance in a data-driven agriculture supply chain. International Journal of Production Economics, 219, 179–194.
Landes, D. S. (1969). The unbound Prometheus. Cambridge University Press.
Lawrence, M., Homer-Dixon, T., et al. (2022). Global polycrisis: The causal mechanisms of crisis entanglement. SSRN 4058592.
Lee, K. (2013). Schumpeterian analysis of economic catch-up. Cambridge University Press.
Manyika, J., et al. (2016). Digital globalization: The new era of global flows. McKinsey Global Institute.
McMillan, M., & Rodrik, D. (2011). Globalization, structural change and productivity growth. NBER Working Paper No. 17143.
Mo Ibrahim Foundation. (2023). 2023 Ibrahim Index of African Governance. Mo Ibrahim Foundation.
Nordling, L. (2017). How the genomics revolution could finally help Africa. Nature, 544(7648), 20–22.
OECD. (2023). OECD Science, Technology and Innovation Outlook 2023. OECD Publishing.

Quadir, I. (2020). South–South cooperation in science, technology and innovation. Technology in Society, 63, 101436.
Rodrik, D. (2013). Unconditional convergence in manufacturing. The Quarterly Journal of Economics, 128(1), 165–204.
Rodrik, D. (2016). Premature deindustrialization. Journal of Economic Growth, 21(1), 1–33.
Rogers, E. M. (2003). Diffusion of innovations (5th ed.). Free Press.
Rotolo, D., Hicks, D., & Martin, B. R. (2015). What is an emerging technology? Research Policy, 44(10), 1827–1843.
Schwab, K. (2016). The fourth industrial revolution. World Economic Forum.
Signé, L., & Johnson, C. (2018). Africa's consumer market potential. Brookings.
Steinmueller, W. E. (2001). ICTs and the possibilities for developing countries to leapfrog. International Labour Review, 140(2), 193–210.
Thornton, P. K., Ericksen, P. J., Herrero, M., & Challinor, A. J. (2014). Climate variability and vulnerability to climate change. Global Change Biology, 20(11), 3313–3328.
Turner, W., et al. (2015). Free and open-access satellite data are key to biodiversity conservation. Biological Conservation, 182, 173–176.
UNCTAD. (2021). Technology and innovation report 2021. United Nations Publications.
UNDESA. (2023). World population prospects 2023. United Nations.
UNDP. (2022). Human Development Report 2021–22. UNDP.
UNESCO. (2021). UNESCO science report. UNESCO Publishing.
Utterback, J. M., & Abernathy, W. J. (1975). A dynamic model of process and product innovation. Omega, 3(6), 639–656.
Varshney, R. K., et al. (2021). Designing future crops: Genomics-assisted breeding comes of age. Trends in Plant Science, 26(6), 631–649.

Vinuesa, R., et al. (2020). The role of artificial intelligence in achieving the Sustainable Development Goals. Nature Communications, 11(1), 1–10.
Wade, R. (1990). Governing the market. Princeton University Press.
World Bank. (2023). Poverty and shared prosperity 2022. World Bank Publications.
World Bank. (2024). World development indicators 2024. World Bank Publications.
WHO. (2021). World Health Statistics 2021. WHO Press.
WIPO. (2023). World Intellectual Property Report 2023. WIPO Publication.

PART II

Emerging Technologies Global Development and Africa's Position

Chapter 3: Classification and Taxonomy

3.1 Introduction

Part II of this volume shifts from the contextual foundations established in Part I to a detailed look at emerging technologies — covering their global status, current adoption in Africa, and potential for development impact. Before exploring individual technologies in Chapter 4, however, it is essential to establish the classification framework, analytical criteria, and enabling conditions that guide the assessment.

A thorough classification and taxonomy serve multiple purposes. They help policymakers prioritize technologies using clear, consistent criteria rather than relying on hype or vendor persuasion. They help researchers identify gaps in evidence and capabilities. Additionally, they provide development practitioners with a systematic approach for matching technologies to specific development challenges.

Various entities have conducted extensive classification efforts, including the Organization for Economic Co-operation and Development (OECD) in its Technology Foresight exercises, the European Commission within its Key Enabling Technologies (KET) framework, the United Nations Conference on Trade and Development (UNCTAD) through its Frontier Technology Readiness Index, and the Asian Development Bank (ADB) for technology assessments in developing Asia. The United States maintains a list of "frontier" or "critical" technologies to guide national security, investment, and research (Office of Science and Technology Policy, 2024).

This chapter adapts and expands these frameworks for the African context, which has a unique mix of urgent development needs, infrastructure challenges, demographic opportunities, and innovation spaces.

3.2 Technology Classification Criteria

Four classification dimensions are used throughout this volume to assess emerging technologies in the African context. These dimensions together create a multi-criteria framework that enables detailed, context-aware technology evaluation—going beyond simple categories to capture the complex trade-offs that decision-makers must navigate.

Maturity Levels: Experimental, Emerging, Maturing, Mature

Technology maturity describes a technology's development and deployment stage. The assessment method used here adapts the Technology Readiness Level (TRL) framework, originally developed by the National Aeronautics and Space Administration (NASA) and later adopted by the European Commission for its Horizon Europe research program (Mankins, 1995; European Commission, 2014). Instead of the original nine-level TRL scale, designed for individual technology components, this version uses a simplified four-stage model suitable for macro-level policy and investment decisions. The four stages are experimental (TR, corresponding to TRL 1–4), emerging (TR, corresponding to TRL 5–6), maturing (TR, corresponding to TRL 7–8), and mature (corresponding to TRL 9–14).

Table 3.1. Technology Maturity Stages: Characteristics, African Examples, and Strategic Implications

Stage	Characteristics	African Examples	Strategy Implication
Experimental	R&D dominant. Technical performance unstable, costs prohibitive, applications limited, extreme uncertainty. TRL 1–3.	Quantum computing (Witswatersrand, Stellenbosch experiments); room-temperature superconductors; general AI.	Monitor through international partnerships. Invest selectively in basic research and scientific collaboration. Avoid large-scale deployment bets.
Emerging	Technical viability demonstrated. Early adopters experimenting. Standards forming. Costs declining but still high. TRL 5–6.	AI-powered diagnostics (Ada Health, Ubenwa); medical drones (Zipline); blockchain supply chains (Bext360 coffee); AR/VR education.	Strategic early investment. Pilot programmes. Regulatory sandboxes. Build absorptive capacity —public–private partnerships for testing.
Maturing	Adoption accelerating. Costs declining rapidly. Ecosystems developing. Competition intensifying. Dominant designs crystallizing. TRL 7–8.	Solar PV; mobile money; cloud computing; IoT for agriculture; e-commerce platforms (Jumia, Takealot); LED lighting.	Rapid scaling. Infrastructure investment. Skills development at scale. Competition policy. Standards adoption. Market incentives.
Mature	Widely diffused. Incremental improvements. Low risk. Established supply chains and maintenance. TRL 9.	4G/5G mobile; basic e-commerce; conventional solar water heating; fibre backbone.	Universal access. Optimization. Transition planning to next generation. Maintenance and sustainability.

Source: Adapted from Mankins (1995) and European Commission (2014)

A key insight about the Technology Readiness Level is that maturity depends on context: a technology may be considered "mature" globally but still be "emerging" in many African countries because of infrastructure gaps, skill shortages, or market conditions. For example, cloud computing is mature in North America and Europe but remains in early stages in Sub-Saharan Africa, where reliable broadband and proximity to data centers are limited. Likewise, precision agriculture is well-established in large-scale farming areas such as the United States, the European Union, Brazil, and Australia, but is still emerging among African smallholders, who need adapted interfaces, affordable sensors, and business models suited to fragmented landholdings. UNCTAD's Frontier Technology Readiness Index (2021) highlights this context-dependent nature by measuring not only the technology's global maturity but also the readiness of specific countries to adopt it.

Development Impact Potential: Transformative, High, Moderate, Low

Development impact potential assesses how effectively a technology can help achieve key development goals for Africa, as outlined in the African Union's Agenda 2063, the Sustainable Development Goals, and national development plans. This evaluation considers not just economic effects but also social, environmental, and institutional factors.

Overall, the development impact of emerging technologies is dynamic rather than fixed. A technology's impact potential changes as enabling conditions shift, costs decrease, and complementary technologies develop. Solar photovoltaics shifted from "low" to "transformative" impact potential within two decades as costs fell by more than 90 percent (International Renewable Energy Agency, 2022). AI-powered diagnostics may follow a similar path as training data for African populations improves and access to computing resources expands. This book, therefore, assesses the impact potential both now and over a 10–15-year period

Table 3.2. Development Impact of Core Emerging Technologies in Alignment with AU Agenda 2063 and SDGs

Impact Level	**Criteria**	**Technology Examples**	**AU Agenda 2063 Sectors**	**SDG Linkages**
Transformative	Potential to fundamentally restructure sectors, create new markets, or shift development trajectories. Broad, multi-sectoral applicability.	AI/ML (cross-cutting); renewable energy systems; mobile connectivity platforms; digital identity infrastructure.	Poverty Reduction; Agriculture & Food Security; Health; Education; Energy; Climate Action; Governance.	SDGs 1, 2, 3, 4, 7, 8, 9, 10, 13, 16, 17.
High	Significant impact in multiple priority sectors. Clear pathways to scale. Demonstrated effectiveness in comparable developing-country contexts.	IoT for agriculture and water; medical drones; blockchain for supply chains; big data analytics; EdTech platforms.	Agriculture & Agro-processing; Health & Nutrition; Education & TVET; Water & Sanitation; Industrial Development.	SDGs 2, 3, 4, 6, 8, 9, 12.
Moderate	Meaningful impact in specific sectors or niches. May require significant enabling conditions. Benefits concentrated rather than broadly distributed.	Industrial robotics; 3D printing; AR/VR for training; nanotechnology applications; autonomous vehicles.	Industry & Manufacturing; Innovation & Technology; Urban Development; Transport; Circular Economy.	SDGs 8, 9, 11, 12.
Low	Limited near-term impact given African contexts. May become relevant as enabling conditions improve or technology matures.	Quantum computing; advanced humanoid robotics; brain–computer interfaces; space-based manufacturing.	Long-term, indirect.	Long-term, indirect.

Source: Author's assessment based on AU Agenda 2063, UN SDGs, and technology deployment data

Risk Profiles: Technical, Social, Environmental, Economic

Emerging technologies pose risks—some inherent to the technology itself, others stemming from their deployment. The risk assessment framework in this book is based on the European Union's AI Act risk classification, the OECD's Principles on AI, and the cautionary frameworks discussed in Chapter 9. This volume assesses four risk dimensions:

- **Technical risks** concern the reliability, security, and performance of the technology itself. These include failure rates and system reliability under African conditions such as heat, dust, humidity, and unstable power supplies; cybersecurity weaknesses in connected systems; interoperability issues between systems from different vendors; and the speed of obsolescence and the availability of maintenance. In the African context, technical risks are often worsened by infrastructure weaknesses—for example, AI systems that rely on dependable connectivity may perform unpredictably in areas with intermittent coverage.

- **Social risks** encompass the impacts of emerging technologies on people, communities, and social structures. These include job displacement, especially in sectors with many low-skilled workers; deepening existing inequalities if technologies mainly benefit the wealthy; loss of privacy and civil liberties through surveillance tools; algorithmic bias that favors certain groups; and disruption of social cohesion through misinformation, polarization, or cultural homogenization. India's experience with Aadhaar—which promoted financial inclusion but also raised serious privacy and exclusion issues—illustrates how complex social risk management can arise in developing-country digital systems (Dreze & Khera, 2017).

- **Environmental risks** relate to the ecological footprint of technology production, deployment, and disposal. These risks include the energy use of digital infrastructure (with data centers accounting for about 1–2 percent of global electricity and their consumption increasing); e-waste from short-lived electronic devices (Africa already bears a disproportionate share of global e-waste); resource extraction for technology hardware (such as cobalt mining in the Democratic Republic of the Congo and lithium in Zimbabwe); and the carbon footprint of manufacturing, transportation, and operations. The European Union's Circular Economy Action Plan and its proposed Ecodesign for Sustainable Products Regulation provide frameworks that African countries can adapt to better manage the environmental impact of technology.

- **Economic risks** include financial sustainability, market dynamics, and macroeconomic effects. These involve high initial costs that may exceed available budgets; vendor lock-in and reliance on single providers; market dominance by major platforms; displacement of local industries by imported technologies; and unsustainable subsidy requirements. Latin America's experience with technology-sector venture capital—a boom-and-bust cycle in 2021–2023 that left many startups underfunded—serves as a cautionary example for Africa's rapidly growing tech investment scene (ECLAC, 2022).

Diffusion Patterns: Adoption Rates and Trends

Technology diffusion—the process by which innovations spread through populations and markets—generally follows predictable patterns but can vary significantly depending on the technology and the context (Rogers, 2003). Understanding these diffusion patterns helps policymakers forecast adoption timelines, identify bottlenecks, and develop strategies to accelerate the adoption of useful technologies while managing disruptive shifts.

Figure 3.1. Speed of Adoption (left) and Stylised S-Curve Trajectories (right)

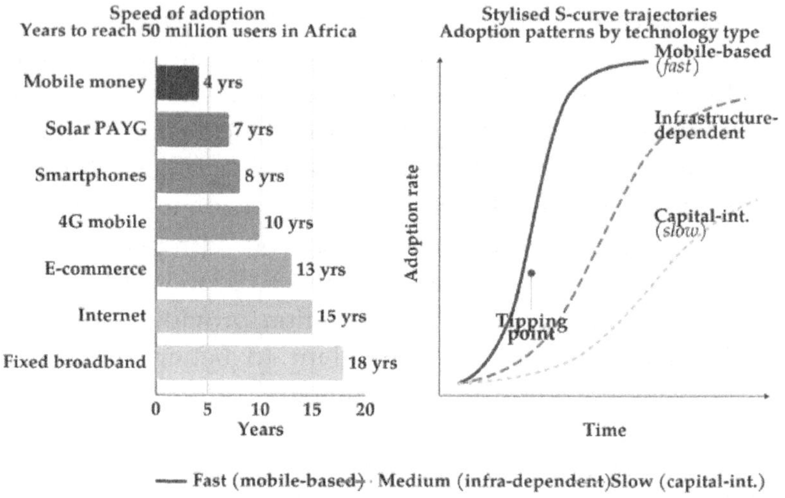

Sources: ITU (2023); GSMA (2023); IRENA (2022); World Bank (2024)

As discussed in Chapter 2, several factors determine the speed of diffusion. Technologies that leverage existing infrastructure (particularly mobile networks) will likely diffuse more quickly than those that require new infrastructure. Technologies with low per-unit costs and clear, immediate user benefits achieve faster adoption.

Regulatory clarity accelerates diffusion; regulatory uncertainty or hostility slows it. Network effects (where value increases with the number of users) create tipping points beyond which adoption accelerates dramatically—as demonstrated by mobile money. Geographic and demographic characteristics matter: densely populated urban areas adopt faster than sparsely populated rural areas, though mobile-based technologies can partially overcome this barrier.

Technology diffusion patterns in Africa are unique in many ways compared to other developing regions, and these differences must be considered.

- First, Africa has demonstrated the capacity for rapid leapfrog adoption when enabling conditions align—mobile money reached significant penetration faster in Africa than comparable digital payment systems in Latin America or South Asia.

- Second, the urban–rural diffusion gap is often wider in Africa due to infrastructure disparities.

- Third, informal market structures and trust networks play a significant role in technology diffusion—agent networks for mobile money and solar home systems rely on social trust rather than formal institutional structures.

- Fourth, regional fragmentation (regulatory, linguistic, and market size) creates adoption discontinuities at borders, slowing cross-country diffusion compared with larger, more integrated markets such as the European Union or China.

3.3 Enabling Layers: Foundations for Technology Ecosystems

Emerging technologies do not operate in isolation. Their adoption depends on a series of enabling layers—spanning physical infrastructure, human capacity, and institutional frameworks—that determine whether deployment is feasible, sustainable, and impactful. This framework builds on established concepts, such as India's digital stack (identity, payments, and data exchange), and the European Commission's Digital Compass, and extends them to the specific development context of Africa. Six enabling layers that are discussed in Chapter 8 are identified: energy systems, connectivity, data infrastructure, compute infrastructure, skills and capabilities, and institutional frameworks.

Figure 3.2. Enabling Layers: Building the Technology Ecosystem Stack

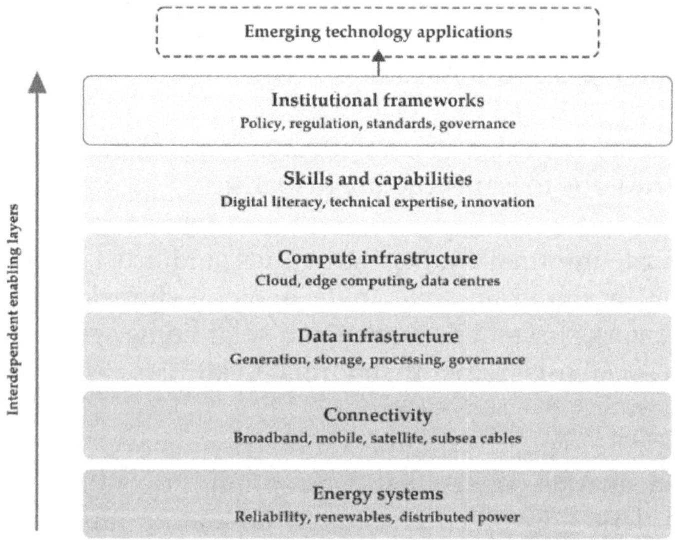

Source: Author's Analysis

Energy Systems

Energy is the most essential enabling layer. Without dependable electricity, no digital technology can operate. Around 600 million Africans—43 percent of the population—lack access to electricity, and many more experience unreliable supply. Power outages are estimated to cost African businesses 2–4 percent of their annual revenue.

The implications of energy systems for technology infrastructure are significant. Data centres require stable, large-scale power; a single hyperscale facility can consume 20–50 MW—comparable to a small city. Mobile towers across the continent frequently rely on expensive, emissions-intensive diesel generators to compensate for grid instability.

Africa's renewable energy endowment offers a potential competitive advantage. The continent holds an estimated 10 TW of solar potential, 350 GW of hydropower, and 110 GW of wind capacity. If harnessed effectively, these resources could power energy-intensive digital services at low cost while attracting sustainability-focused investment from global cloud providers.

Connectivity

Connectivity is the circulatory system of the digital economy. As of 2023, only 37 percent of Africans use the Internet, compared with a global average of 66 percent and 68 percent in Latin America and the Caribbean. The urban–rural divide is pronounced: penetration in African cities often exceeds 50 percent while rural rates fall below 10 percent.

Africa's connectivity ecosystem is nonetheless evolving rapidly across multiple layers. Major submarine cable projects—including 2Africa (the world's longest at 45,000 km, completed in 2025), Equiano (Google, operational in 2023), and PEACE (operational in 2025)—have significantly increased international bandwidth. By 2025, 4G coverage reached roughly 60 percent of the population, though rural coverage lags. Low Earth orbit (LEO) satellite constellations such as Starlink, OneWeb, and Amazon's Project Kuiper offer new opportunities to extend broadband to areas unserved by terrestrial networks, though affordability remains a barrier.

Data Infrastructure

Data is the fundamental input for nearly all emerging technologies—AI, IoT, blockchain, precision agriculture, and health tech alike. The data infrastructure layer encompasses the full value chain: generation, storage, processing, and governance.

Africa faces significant gaps at each stage. Low digitisation of public services, minimal sensor deployment, and weak national statistical systems constrain data generation. Storage is scarce: Datacentermap indicates that as of 2025, Africa hosts only 268 of more than 11,000 data centres worldwide (2.5 percent), despite representing 17 percent of the global population. Processing capacity for AI workloads is largely located outside the continent, raising both latency and sovereignty concerns.

Governance frameworks are improving. The African Union Data Policy Framework (2022) provides a continental reference point, and many countries are establishing Data Protection Authorities. International models such as India's Aadhaar identity platform and the Unified Payments Interface demonstrate how large-scale digital public infrastructure can be built in developing contexts, while the EU's GDPR-centred approach emphasises rights protection alongside infrastructure investment. Africa's challenge is to build a high-quality, accessible data infrastructure that serves development needs without falling into data colonialism (unchecked extraction) or data isolation (excessive localisation).

Compute Infrastructure

Compute capacity underpins cloud services, AI model training, application hosting, and IoT data processing. Global compute demand is growing exponentially—training a large language model in 2023 required approximately 10,000 times as much compute as in 2018.

Africa's compute infrastructure is growing, with major cloud providers (AWS, Microsoft Azure, Google Cloud, and Huawei Cloud) establishing or announcing their presence in South Africa, Kenya, Nigeria, and Egypt. Yet, many African countries still lack local cloud infrastructure, making them reliant on distant data centers with higher latency and greater data sovereignty risks. Edge computing—processing data near its source—offers significant potential for the continent by reducing dependence on centralized infrastructure and supporting applications in low-connectivity areas.

Skills and Capabilities

Human capital is the most critical and yet the most difficult enabling layer to develop. Technology ecosystems require capabilities at various levels: population-wide digital literacy, technical skills (programming, data science, cybersecurity), specialised AI expertise, entrepreneurial capacity, and institutional skills for policy and regulation.

Africa's skills base remains constrained. In 2021, Sub-Saharan Africa produced approximately 8 percent of its tertiary graduates in STEM fields, compared with 22 percent in East Asia and 24 percent in Europe. The continent has roughly 96 researchers per million people, against 1,340 in East Asia and 4,190 in Europe. Brain drain exacerbates the gap.

Encouraging counterforces are at work. The African diaspora contributes through remittances, knowledge networks, and return migration. A growing ecosystem of coding bootcamps and accelerators—including Andela, ALX Africa, Moringa School, Semicolon, WeThinkCode, and Holberton School Africa —is training tens of thousands of developers and data scientists across the continent.

Interdependencies and System Requirements

The six enabling layers—data infrastructure, compute, connectivity, energy, skills, and institutions—are deeply interdependent. Weakness in any single layer constrains the entire system, creating bottlenecks that limit technology adoption, even as progress occurs in other layers. For example, AI deployment requires not only algorithms (technology) but also training data (data layer), processing power (compute), Internet access for cloud services (connectivity), reliable electricity (energy), skilled practitioners (skills), and proper governance frameworks (institutions). Achieving real impact requires coordinated investment across all layers, rather than addressing each layer sequentially.

This system's perspective has important implications for emerging technology strategy. Countries that invest heavily in one layer while ignoring others may not see proportional returns. Conversely, modest, coordinated improvements across multiple layers can lead to significant gains.

Rwanda's success in digital transformation illustrates this principle: simultaneous investment in fibre infrastructure, regulatory frameworks, digital skills programmes, e-government platforms, and strategic partnerships created a mutually reinforcing system greater than the sum of its parts. India's digital public infrastructure strategy similarly succeeded by coordinating identity (Aadhaar), payments (UPI), and data exchange (DEPA) layers. Countries at earlier stages of digital development need to identify the key constraints across the stack and address them in parallel rather than in isolation.

3.4 Global Technology Innovation Clusters and Africa's Participation

Global innovation is highly concentrated. The United States, China, the United Kingdom, European Union, Japan, and the Republic of Korea together account for approximately 89 percent of global R&D spending, over 90 percent of patent applications, and most of the world's tech companies by market capitalization (OECD, 2023; World Intellectual Property Organization [WIPO], 2023).

Figure 3.3. Concentration of Global Research and Development Expenditure by Region (2023)

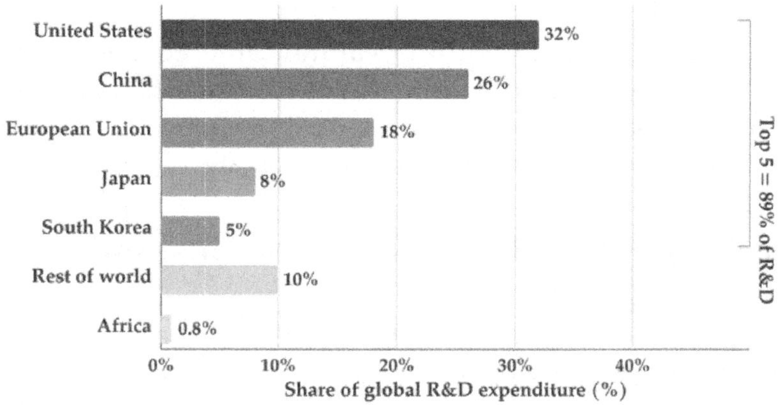

Sources: UNESCO (2021); OECD (2023)

Within these regions, innovation hubs tend to cluster in specific metropolitan areas: Silicon Valley, Boston, Shenzhen, Beijing, Seoul, Tel Aviv, London, and Stockholm, among others, benefiting from agglomeration effects that reinforce their advantages.

This concentration has important implications for Africa.

- First, the technologies that will shape African development are mostly created and designed elsewhere, often without considering African contexts—AI training data underrepresents African languages, faces, and environments; hardware is designed for temperate climates and reliable power supplies; and user interfaces assume literacy levels and cultural norms that may not apply.

- Second, intellectual property regimes and technology licensing terms are controlled by actors in major innovation centers, often creating barriers for African adopters.

- Third, global venture capital and corporate R&D investments tend to flow mainly to established innovation centers, making it hard for African innovators to secure the capital they need to compete at scale.

Leading Firms, Research Institutions, and Ecosystems

The global technology landscape is dominated by a small number of very large firms. The 'Big Tech' companies—Alphabet (Google), Amazon, Apple, Meta, Microsoft, and increasingly Chinese firms such as Huawei, Tencent, Alibaba, and ByteDance—possess vast resources, talent, data assets, and market influence. Their combined R&D spending surpasses that of most countries. These firms increasingly shape African digital ecosystems through cloud platforms, social media, search engines, app stores, and hardware—especially smartphones. Their decisions regarding African-language support, data center locations, content moderation, and pricing significantly impact African users and developers.

Research institutions leading the development of frontier technology are also based in the top countries. These include universities (Massachusetts Institute of Technology, Stanford University, Carnegie Mellon University, University of Oxford, University of Cambridge, University of Toronto, University of Alberta, ETH Zurich, Max Planck Institute, Tsinghua University, Peking University), government research labs (United States National Institutes of Health, European Organization for Nuclear Research (CERN), Chinese Academy of Sciences), and public–private research partnerships (Fraunhofer Institutes in Germany, RIKEN in Japan, CSIR in India).

Africa's leading research institutions, like the University of Cape Town, the University of the Witwatersrand, the University of Lagos, Makerere University, the University of Nairobi, and Cairo University, have growing capabilities but face resource constraints that limit competitiveness in capital-intensive research.

The low level of participation in research and development has wide-ranging implications. Africa's position in global technology value chains is primarily as a consumer of finished products and a provider of raw materials. The continent imports over 90 percent of its technology hardware—such as smartphones, computers, servers, network equipment, solar panels, and electronic components—with limited local assembly or manufacturing.

Conversely, Africa supplies critical minerals vital to technology hardware: the Democratic Republic of the Congo produces about 70 percent of global cobalt (essential for batteries); South Africa dominates platinum-group metals and manganese; Zimbabwe, Mali, and Namibia have substantial lithium reserves; and other countries hold deposits of rare-earth minerals. The challenge for Africa is to move beyond just raw-material extraction and focus on local processing, component manufacturing, and value addition—a goal that requires industrial policy, technology transfer, and significant investments in infrastructure and skills.

Technology Transfer Mechanism Potentials and Barriers

Technology transfer—the process by which technologies developed in one context are adopted and adapted in another—is critical to African technology development. Transfer occurs through multiple channels: foreign direct investment (FDI) by technology companies, licensing agreements, joint ventures, imports of capital goods, diaspora knowledge networks, open-source software and hardware, development assistance, and South–South cooperation. Africa could benefit greatly from technology transfer, yet each channel has distinct advantages and limitations.

Barriers to technology transfer include intellectual property protections that restrict access, high licensing fees, conditions imposed by technology providers (such as data residency requirements and proprietary standards), limitations in absorptive capacity (like insufficient skills to understand, adapt, and maintain transferred technologies), and geopolitical factors (for example, export controls on advanced semiconductors, which mainly target China but affect global supply chains).

The EU–African Union Strategic Partnership, China's Belt and Road Initiative digital dimension, India's Lines of Credit programme, Japan's Tokyo International Conference on African Development (TICAD) process, and the Republic of Korea's Knowledge Sharing Programme all include technology transfer components, with varying levels of effectiveness and conditions. In Africa, the strategic challenge is to maximize technology transfer while building indigenous capabilities that reduce long-term dependence.

Table 3.3. Advantages and Limitations of Different Technology Transfer Channels

Transfer Channel	Advantages	Limitations
FDI by Technology Companies	Capital injection at scale; rapid diffusion of operational practices; managerial learning; integration into global value chains; supplier spillovers.	Enclave development risk; limited IP transfer; profit repatriation; digital dependency; weak spillovers.
Licensing Agreements	Access to proven technology; faster diffusion; builds domestic production capacity; useful in pharma, telecom, software.	High royalties; limited know-how; restrictive clauses; IP dependency.
Joint Ventures	Shared ownership; incentivized knowledge transfer; skills upgrading; balances foreign capital and domestic control.	Power asymmetries; governance conflicts; core tech retained by foreign partner.
Imports of Capital Goods	Quick modernization; access to advanced equipment; supports Industry 4.0 transition.	Limited tacit transfer; maintenance dependency; foreign exchange burden; trade deficits.
Diaspora Knowledge Networks	Tacit knowledge transfer; global innovation linkages; startup ecosystem support.	Fragmented; limited scale; brain drain risks; institutional absorption constraints.
Open-Source Software & Hardware	Low-cost access; customization; reduced dependency; enables AI & IoT innovation.	Requires high skills; sustainability challenges; Global North governance dominance.
Development Assistance	Infrastructure support; regulatory reform; capacity building; risk reduction.	Donor-driven priorities; fragmentation; sustainability challenges.
South–South Cooperation	Context-relevant models; infrastructure acceleration; emerging tech partnerships.	Geopolitical dependency risks; debt risks; transparency concerns.

Source: Author's compilation based on UNCTAD (2021); World Bank (2024); African Development Bank (2023)

3.5 Africa's Readiness and Adoption Landscape

Continental Readiness Assessment Frameworks

Continental readiness is crucial for technology transfer and maintaining competitiveness. Several international frameworks assess technology readiness in developing countries. The most relevant include UNCTAD's Frontier Technology Readiness Index, which assesses ICT, skills, R&D, industry, and access to finance; the World Economic Forum's Networked Readiness Index, covering technology, people, governance, and impact; the ITU's ICT Development Index (IDI), measuring access, use, and skills; the Global Innovation Index (GII), published by WIPO, which evaluates innovation inputs and outputs; and the AfDB's Africa Infrastructure Development Index (AIDI), focusing on physical and digital infrastructure.

These indices show a consistent trend: African countries generally score below global averages on technology readiness, but with wide variation. A small number of "champion" countries—Mauritius, South Africa, Kenya, Rwanda, Egypt, Morocco, Tunisia, Botswana, and Ghana—score well above the continental average and compete with middle-income countries in other regions. A larger group of countries is making progress but faces notable gaps. Some countries—especially those impacted by conflict, extreme poverty, or geographic isolation—have limited readiness to adopt frontier technologies and require foundational investments before advanced technologies become relevant.

The comparative data in Table 3.4 show a significant gap in Africa across all measured aspects of technology readiness. However, two nuances are important. First, Africa's starting point is lower, but its rate of progress across several indicators (especially mobile connectivity, renewable energy, and fintech) has been among the fastest globally. Second, Africa's readiness gap is not uniform across the continent. Some African countries score as high as, or higher than, those in Asia and Latin America on certain indicators, indicating that targeted, well-planned efforts can lead to rapid progress.

Table 3.4. Technology Readiness Indicators: Regional Comparison (2023)

Indicator	Sub-Saharan Africa	South Asia	Latin America & Caribbean	East Asia & Pacific
Internet penetration	37%	43%	68%	70%
Mobile broadband (3G+)	50%	65%	82%	90%
Electricity access	57%	96%	98%	98%
R&D (% of GDP)	0.4%	0.6%	0.7%	2.4%
Researchers/million	96	257	489	1,340
STEM grads (% of total)	8%	16%	13%	22%
VC investment (2023)	$3.2B	$9.8B	$8.5B	$72B
Data centres	<100	~150	~250	>600

Sources: ITU (2023); UNESCO (2021); IEA (2022); World Bank (2024); GSMA (2023); Partech Africa (2024)

The most important finding from readiness assessments is the significant diversity across African countries. Country readiness profiles vary greatly depending on factors such as economic structure and income levels; institutional quality and governance capacity; infrastructure, including energy, connectivity, and transportation; human capital development; the maturity of innovation ecosystems; and strategic policy decisions on technology priorities.

Table 3.5. Technology Readiness Profiles: Selected African Countries

Country	Key Strengths	Key Gaps	Distinctive Niche
South Africa	Strongest R&D and university system. Advanced financial sector. Largest data-centre market on the continent.	High inequality limits broad adoption. Energy crisis (load-shedding). Skills mismatches. Political uncertainty.	Research hub; automotive and mining tech; renewable energy (REIPPPP); financial innovation.
Kenya	Mobile money pioneer (M-PESA). Vibrant startup ecosystem ('Silicon Savannah'). Regulatory innovation in fintech.	Infrastructure gaps outside Nairobi. High youth unemployment. Data governance still evolving.	Fintech; AgTech; digital services export; geothermal energy.
Nigeria	Largest market (220M population). Dynamic fintech sector ($1.3B+ VC in 2022). Creative industries (Nollywood, Afrobeats).	Severe infrastructure deficits (power, broadband). Regulatory inconsistency. Security challenges in some regions.	Fintech; e-commerce; creative economy; market scale.
Rwanda	Strong governance and policy coherence. Drone delivery leader (Zipline). Digital transformation strategy. Regulation.	Small market (13M). Limited private sector depth. Low income constraining demand. Landlocked geography.	Digital governance; drone logistics; ICT hub; smart city (Kigali).
Morocco	Strategic location. Industrial policy focus (automotive, aerospace). Solar leader (Noor-Ouarzazate). EU proximity.	Urban–rural divide. Youth unemployment. Limited VC ecosystem.	Renewable energy; automotive manufacturing; fintech gateway to Francophone Africa.
Ghana	Stable democracy. Drone regulatory framework leader. Growing tech hubs (Accra). Blockchain for land registry.	Fiscal pressures. Power-sector challenges. Brain drain.	Drone regulation model; digital governance; cocoa value-chain tech.
Senegal	Francophone hub. DigiWorld Dakar. Growing startup ecosystem. Mobile money growth.	Limited broadband outside Dakar. Low electricity access in rural areas. Small market.	Francophone tech hub; digital agriculture; gateway to West Africa.

Source: Author's assessment based on UNCTAD (2021); WIPO (2023); World Bank (2024); national development plans.

The variation has significant implications for the adoption of emerging technologies across Africa. It indicates that no single technology strategy or deployment model can be applied uniformly throughout the continent. Countries with more developed digital ecosystems are better positioned to test and scale emerging technologies, including AI, IoT, data analytics, and blockchain. Meanwhile, countries just starting their digital journey need to address fundamental gaps in energy access, connectivity, and digital literacy to implement advanced technologies effectively.

This heterogeneity also creates opportunities for intra-African knowledge exchange: countries that have successfully tackled specific challenges—such as Rwanda in digital governance, Kenya in mobile financial services, and Morocco in industrial technology integration—can serve as models and mentors for peers facing similar issues.

Regional economic communities and the African Continental Free Trade Area can play vital roles in facilitating this exchange and harmonizing standards to support cross-border diffusion of technology. Ultimately, the remarkable diversity of readiness profiles across Africa's countries requires that technology adoption strategies be tailored to local contexts—building on existing strengths, addressing the most critical constraints, and leveraging proven African innovations that have already gained global recognition.

3.6 Conclusion

This chapter establishes the classification framework, enabling conditions, and readiness landscape that shape the analysis of specific emerging technologies in Chapter 4. The four classification dimensions—maturity, development impact potential, risk profile, and diffusion patterns—provide a systematic, transparent basis for evaluating technology. The six enabling layers—data and compute infrastructure, connectivity, energy, skills, and institutions—define the ecosystem requirements that determine whether technology adoption is feasible and sustainable. The mapping of global innovation leaders highlights both the structural challenges Africa faces as a peripheral player in global innovation systems and the specific niches where African innovation is already gaining global significance. The readiness assessment reveals both Africa's overall gap compared to other regions and the notable variability across African countries, indicating that technology strategies need to be tailored rather than taken as one-size-fits-all.

Chapter 4 expands on this foundation by exploring each frontier technology in detail: its global state of development, adoption and use cases in Africa, key players and markets, growth potential, and associated risks.

References

Africa: The Big Deal. (2024). African startup funding report 2024. Africa: The Big Deal.

African Development Bank. (2023). African economic outlook 2023: Mobilizing private sector financing for climate and green growth in Africa. African Development Bank Group.

Christensen, C. M. (1997). The innovator's dilemma: When new technologies cause great firms to fail. Harvard Business School Press.

Dreze, J., & Khera, R. (2017). Recent social security initiatives in India. World Development, 98, 555–572.

Economic Commission for Latin America and the Caribbean. (2022). Digital technologies for a new future. ECLAC.

Epoch AI. (2023). Trends in machine learning compute. Epoch AI.

European Commission. (2014). Technology readiness levels (TRL): Horizon 2020 — Work programme 2014–2015. European Commission.

European Commission. (2021). 2030 Digital Compass: The European way for the Digital Decade. European Commission.

GSMA. (2023). State of the industry report on mobile money 2023. GSMA.

GSMA. (2023). The mobile economy: Sub-Saharan Africa 2023. GSMA.

International Energy Agency. (2022). Africa Energy Outlook 2022. IEA Publications.

International Renewable Energy Agency. (2022). Renewable capacity statistics 2022. IRENA.

International Telecommunication Union. (2023). Measuring digital development: Facts and figures 2023. ITU Publications.

Mankins, J. C. (1995). Technology readiness levels: A white paper. NASA Office of Space Access and Technology.

Organization for Economic Co-operation and Development. (2023). OECD Science, Technology and Innovation Outlook 2023. OECD Publishing.

Partech Africa. (2024). Africa Tech Venture Capital Report 2024. Partech Partners.
Privacy International. (2023). State of privacy: Africa. Privacy International.
Rogers, E. M. (2003). Diffusion of innovations (5th ed.). Free Press.
UNESCO. (2021). UNESCO science report: The race against time for smarter development. UNESCO Publishing.
United Nations Conference on Trade and Development. (2021). Technology and innovation report 2021: Catching technological waves—Innovation with equity. United Nations Publications.
United States Office of Science and Technology Policy (2024). Critical and Emerging Technologies List Update. Government Publications.
World Bank. (2024). World development indicators 2024. World Bank Publications.
World Economic Forum. (2023). Network readiness index 2023. World Economic Forum.
World Intellectual Property Organization. (2023). Global Innovation Index 2023. WIPO Publication.

Chapter 4: Emerging Technologies—Global Trends and African Experiences

4.1 Introduction

This chapter forms the core of the volume's technical content, offering structured evaluations of 10 emerging technologies with the potential to transform African development. Each technology is examined using a uniform framework: definition and core concepts; global state of the art; Africa's current adoption and use cases; key actors and markets; development potential and opportunities; and risks and challenges. This consistency allows for cross-technology comparisons and supports the classification framework introduced in Chapter 3.

The technologies cover the full range of maturity and impact: from established ones already scaling in Africa (renewable energy systems, mobile-enabled big data analytics) to emerging technologies showing early but promising deployment (AI diagnostics, drone delivery, blockchain supply chains), and experimental technologies at the forefront of global research with long-term relevance for Africa (advanced robotics). Together, they form the technology portfolio that will shape Africa's development over the coming decades.

Throughout the chapter, global comparative insights are included. African examples cover all regions—North, West,

East, Central, and Southern Africa—to highlight the diversity of innovation across the continent.

Africa's best strategy is to prioritize deployable, inclusive technologies (renewables, IoT, drones) while strategically building governance, skills, and infrastructure capacity to leverage high-impact frontier technologies like AI, blockchain, IoT and robotics.

- **Renewable energy** stands out as the most development-oriented and socially inclusive technology, offering significant job-creation potential, high social impact, and relatively low systemic risks — making it fundamental to structural transformation.

- **AI** has the greatest projected impact on development and attracts the strongest investment flows. However, it also presents higher risk profiles. AI is less developed in most African countries, highlighting the need for stronger governance capacity, skills development, and regulatory frameworks to address inequality and concentration risks.

- **IoT and drones** occupy a practical middle ground, offering significant development and inclusion benefits with moderate risks, making them transitional technologies capable of delivering near-term gains across sectors such as agriculture, healthcare, logistics, and smart infrastructure.

- **Blockchain** demonstrates notable potential for inclusion —particularly in financial systems and digital identity— though its investment depth and maturity in Africa remain moderate.

- **Robotics** appears the least mature and most capital-intensive, suggesting it has longer-term, niche relevance rather than immediate transformational impact.

Figure 4.1. Comparative Technology Assessment for African Development

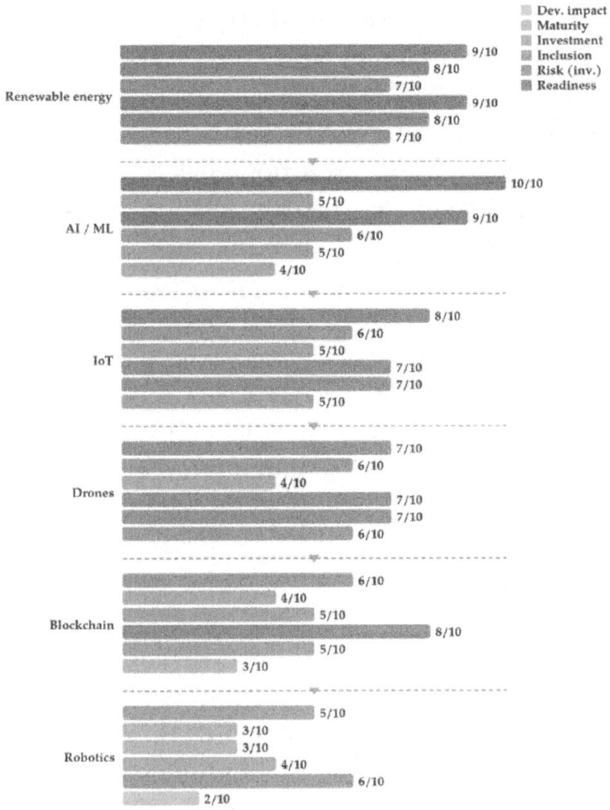

Source: Author's Analysis.

4.2 Additive Manufacturing (3D Printing)

Definition and Core Concepts

Additive manufacturing (AM), often called 3D printing, produces three-dimensional objects by layering material from a digital design file.

The technology covers various processes:

- **Fused deposition modelling (FDM)** extrudes thermoplastic filament and is the most affordable and accessible method.
- **Stereolithography (SLA)** uses ultraviolet light to cure liquid resin for high-resolution objects.
- **Selective Laser Sintering (SLS)** fuses powdered materials using a laser to produce functional engineering parts.
- **Metal printing** (direct metal laser sintering, electron beam melting) produces aerospace-grade components and surgical implants.
- **Bioprinting** deposits living cells to create tissue constructs for medical research (Ngo et al., 2018; Gibson et al., 2015).

Global State of the Art

The global additive manufacturing market was valued at approximately $18 billion in 2023, with projections reaching $65 billion by 2030 (Wohlers Associates, 2023). The industry has evolved from rapid prototyping to producing end-use parts on an industrial scale—aerospace companies such as Boeing, Airbus, and GE Aviation print certified flight components. Medical device manufacturers create patient-specific surgical guides, implants, and dental prosthetics. Construction firms in the United States, China, the Netherlands, and the United Arab Emirates have showcased 3D-printed houses and offices. Europe's Horizon Europe program and Germany's Fraunhofer Institutes have positioned the region as a leader in additive manufacturing, competing with the United States and China.

Africa's Current Adoption and Use Cases

Africa's adoption of additive manufacturing is mainly in South Africa, Kenya, Nigeria, Egypt, and Morocco. South Africa leads with the Council for Scientific and Industrial Research (CSIR) National Laser Centre, which operates metal additive manufacturing systems, and the Central University of Technology's Centre for Rapid Prototyping and Manufacturing (CRPM), which produces patient-specific titanium implants.

Organizations like Not Impossible Labs have created low-cost 3D-printed prosthetic hands for amputees in Sudan, Kenya, and Uganda, costing $50–100 compared to $5,000–50,000 for traditional prosthetics (Not Impossible Labs, 2020). In construction, 14Trees built Africa's first 3D-printed house in Malawi in 2021, taking 12 hours (14Trees, 2021). Kenyan maker spaces (Gearbox, Nairobi) and Nigerian fabrication labs are expanding local manufacturing skills.

Figure 4.2. Additive Manufacturing in Africa: Current Adoption vs. Future Potential

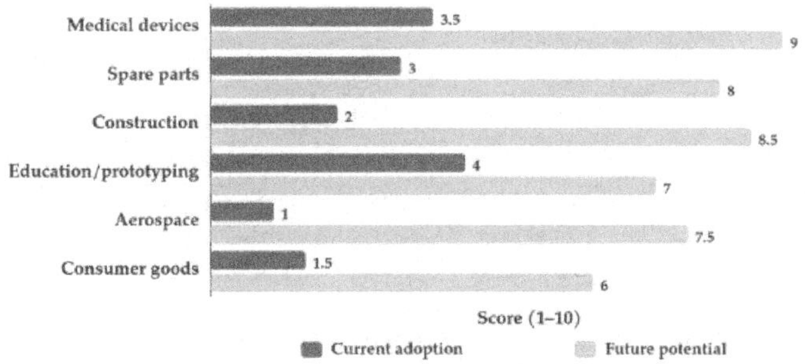

Source: Author's Analysis

Development Potential

Additive manufacturing offers strategic advantages for Africa, especially in replacing imports and boosting local production. Many African economies rely heavily on imported manufactured goods, such as spare parts, medical supplies, and industrial components, making them susceptible to supply chain disruptions and foreign exchange issues. The COVID-19 pandemic exposed this vulnerability and revealed opportunities for local fabrication, as makerspaces and universities across Africa produced face shields, ventilator parts, and other essential items when global supply chains broke down (AU-NEPAD, 2021).

Additive manufacturing promotes decentralized, small-batch production, which can reduce downtime in sectors such as mining, agriculture, and transportation. It also helps strengthen regional value chains within the African Continental Free Trade Area. Over time, local digital manufacturing networks could enhance industrial resilience and help address structural trade imbalances.

In healthcare, additive manufacturing provides transformative opportunities through patient-specific solutions and local device production. Africa continues to face shortages of prosthetics, implants, and specialized medical devices, often depending on expensive imports with extended delivery times. Additive manufacturing allows for the creation of customized prosthetics, orthopedic implants, dental devices, and anatomical surgical models tailored to individual patients. This helps reduce costs, shortens waiting times, and improves clinical outcomes.

The construction industry is another high-impact sector, especially considering Africa's estimated housing shortfall of 51 million units (African Development Bank, 2023). Rapid urbanization and population growth have outpaced traditional building methods, which are often labor-intensive, slow, and wasteful. Large-scale 3D printing technologies could accelerate housing delivery, reduce material waste, and support environmentally friendly designs.

Education and capacity-building in innovation could be the most significant impact of additive manufacturing. Affordable desktop 3D printers in schools, technical institutes, universities, and innovation hubs promote hands-on learning and rapid prototyping, improving design thinking and engineering skills. By enabling students and entrepreneurs to quickly transform ideas into physical models, AM promotes a culture of hardware entrepreneurship and problem-solving.

Taken together, these four areas show that additive manufacturing is more than just a production method; it is a distributed industrial capability platform. The main policy challenge is developing the ecosystem—skills training, regulatory systems, material supply chains, financing options, and alignment with continental trade strategies. If integrated into national industrial plans and Agenda 2063 priorities, additive manufacturing can shift from a niche to a central element in Africa's future manufacturing growth.

Risks and Challenges

The adoption of additive manufacturing in Africa faces major structural and economic challenges, starting with the high cost of industrial-grade equipment. While desktop 3D printers are becoming more affordable, advanced metal or large-scale industrial systems suitable for aerospace, medical, or construction applications typically cost between $200,000 and $1 million, making them out of reach for most small and medium-sized enterprises and technical schools. Limited access to financing and high borrowing rates also slow the adoption of capital-intensive technologies. Furthermore, specialized materials—such as high-quality polymers, metal powders, and biocompatible materials—are often scarce in Africa, resulting in increased imports and higher production costs.

A second set of challenges relates to skills and institutional capacity. Additive manufacturing requires advanced design skills, including computer-aided design (CAD), materials science expertise, machine calibration, and post-processing.

Intellectual property (IP) concerns add further complexity, especially when design files are easily shared or copied.

Quality assurance and regulatory frameworks pose additional challenges, especially in the medical and structural fields, where safety standards are crucial. In healthcare, 3D-printed implants and devices must comply with strict certification requirements to ensure patient safety, but many African regulatory agencies lack the necessary technical expertise and testing facilities to evaluate these products. The European Union's CEN/CENELEC standardization approach provides a useful model for Africa, demonstrating how regional standards organizations can unify quality, safety, and technical standards across markets.

4.3 Augmented Reality (AR)

Definition and Core Concepts

Augmented reality overlays digital information—such as images, text, 3D models, and animations—onto the user's view of the physical world, enhancing perception and interaction. AR relies on hardware (cameras, displays, sensors) and software (computer vision, spatial mapping) to accurately align digital content with the physical environment in real time (Azuma, 1997). Smartphone-based AR is the most relevant platform in Africa, leveraging devices users already own without the need for expensive dedicated hardware.

Global State of the Art

The global AR market is expanding rapidly, demonstrating technological maturity and increasing enterprise adoption. Valued at around $32 billion in 2023, it is expected to reach over $90 billion by 2028 (Grand View Research, 2024), driven by advances in spatial computing, wearable devices, edge processing, and AI integration. Major tech companies are making AR a key part of next-generation human–computer interaction. Currently, enterprise use is driving AR's growth—applications such as remote assistance, immersive training simulations, collaborative prototyping, and logistics navigation demonstrate AR's economic value beyond entertainment.

Africa's Current Adoption and Use Cases

African AR adoption is still emerging but expanding through smartphone applications. Educational uses include AR chemistry and biology tools in South African schools and AR-integrated textbooks in Kenya. The African Virtual University explores the use of AR for laboratory simulations. Tourism and cultural heritage projects add historical information to sites in Egypt, Ethiopia, Zimbabwe, and South Africa. Nigerian developers have built AR applications for Yoruba cultural education. Field services in South African mining and telecommunications use AR-guided remote assistance.

Development Potential

Augmented reality has significant growth potential in Africa, especially in bridging skills gaps and expanding access to technical training. AR-enabled immersive training tools can simulate complex procedures—from equipment maintenance to medical treatments—without the need for expensive physical setups. In agriculture and healthcare, AR can facilitate real-time knowledge sharing and service delivery, particularly in rural and underserved areas. Additionally, AR can support industrial modernization and foster the development of digital entrepreneurial ecosystems.

Risks and Challenges

Deploying augmented reality across Africa faces structural and ecosystem-level challenges that extend beyond device availability. Many low-end smartphones commonly used in

African markets lack sufficient RAM, processing power, and battery life to support seamless AR experiences. Connectivity also presents a significant challenge—real-time AR overlays depend on high-speed 4G or 5G networks with low latency. Content localization is another significant challenge, as many AR platforms do not support African languages, cultural stories, indigenous knowledge, or locally relevant data.

4.4 Virtual Reality (VR)

Definition and Core Concepts

Virtual reality creates fully immersive, computer-generated environments experienced through head-mounted displays that track head movements and adjust the visual perspective. VR may include spatial audio, haptic feedback, and motion controllers. The range varies from 360-degree videos on smartphone-based headsets to fully interactive worlds that require powerful computing hardware (Jerald, 2015).

Global State of the Art and African Applications

In 2023, the VR market was valued at approximately $28 billion, driven by rapid advancements in hardware, content ecosystems, and enterprise adoption (Statista, 2024). Meta has invested over $40 billion since 2019 in metaverse-related infrastructure and platforms. In healthcare, VR-enabled surgical simulation and anatomy visualization improve procedural accuracy and lower training risks.

In Africa, specific applications demonstrate strong potential—such as medical training at the University of Cape Town, mining safety simulations in South Africa, architectural visualization in Nigeria and Kenya, and VR-based therapies for Post-Traumatic Stress Disorder.

Risks and Challenges

Hardware costs are the primary immediate barrier to VR adoption. High-quality head-mounted displays, motion controllers, and compatible computing devices are often too expensive for most households, schools, and small businesses. Entry-level standalone devices (e.g., Meta Quest) can cost several hundred dollars, while high-end VR systems require gaming PCs costing over $1,000. In schools or training centers, scaling VR requires purchasing multiple devices, which significantly raises upfront capital costs and ongoing maintenance expenses. Even when devices are affordable, reliable electricity and connectivity are often unavailable.

Further, VR systems are increasingly collecting sensitive biometric data—including eye-tracking, gesture patterns, and physiological responses—raising serious concerns about data privacy and cybersecurity. However, adaptive models provide a solution—shared VR facilities at universities and TVET centers can help distribute hardware costs among multiple users.

4.5 Autonomous Vehicles and Drones

Definition and Core Concepts

Unmanned Aerial Vehicles are cyber-physical systems capable of sensing their environment, making decisions, and navigating with little or no human assistance. These systems integrate sensing, computing, control, and actuation in real time. Drones range from remotely piloted aircraft systems (RPAS), in which human operators maintain active control, to fully autonomous platforms that execute pre-programmed missions or adapt to changing conditions using onboard intelligence (Valavanis & Vachtsevanos, 2015). For ground autonomous vehicles, the Society of Automotive Engineers (SAE) classifies driving automation into six levels, from Level 0 (no automation) to Level 5 (full automation under all conditions).

Global State of the Art

The global drone industry has grown rapidly from specialized military and hobbyist uses to a large commercial market, valued at about $35 billion in 2023 and expected to surpass $90 billion by 2030 (Drone Industry Insights, 2024). In logistics, companies like Amazon and Wing have been testing drone-based last-mile delivery.

In agriculture, China has deployed over 200,000 agricultural drones for crop spraying and precision farming. Autonomous ground vehicles have also moved from experimental prototypes to early-stage commercial deployment—Waymo runs commercial robotaxi services in select U.S. cities. The European Union's U-space framework creates a harmonized drone traffic management system.

Africa's Current Adoption and Use Cases

Africa has emerged as a global leader in the use of autonomous aerial systems for development, particularly in public health logistics. Rwanda and Ghana's partnerships with Zipline are among the most successful large-scale deployments of autonomous drone services for health worldwide. Since 2016, Zipline's fixed-wing drones have made over 10 million deliveries of vaccines, blood products, and vital medical supplies to remote health facilities (Zipline, 2024). Zipline has also expanded into Nigeria, Kenya, and Côte d'Ivoire.

Figure 4.3. Autonomous Vehicles and Drones in Africa

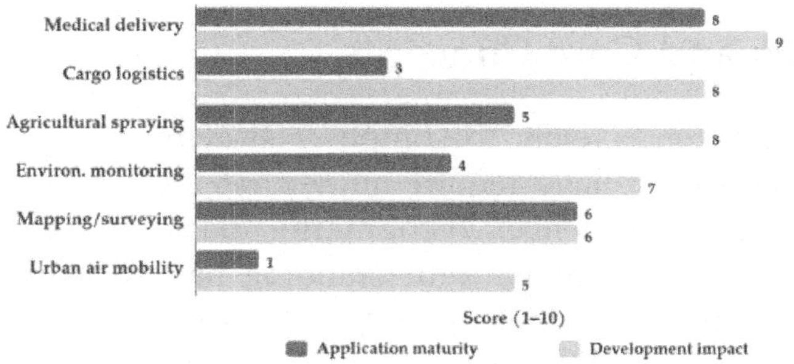

Source: Author's Analysis

The Ghana Civil Aviation Authority developed one of the world's first comprehensive regulatory frameworks for commercial drone operations. Besides healthcare, autonomous aerial systems are increasingly used in agriculture, urban planning, environmental monitoring, and land management.

Risks and Challenges

The deployment of drones and autonomous vehicles across Africa faces complex regulatory, technical, and socio-economic challenges. Regulatory gaps remain significant in many countries, especially regarding beyond-visual-line-of-sight (BVLOS) operations, airspace integration, liability distribution, and cross-border harmonization. Privacy concerns are rising as drones equipped with high-resolution cameras and AI-enabled analytics can collect detailed personal data. Safety risks—including mid-air collisions, system failures, and cybersecurity breaches—require strong certification and monitoring systems. Proactive governance models in Ghana and Rwanda show that adaptive, risk-based oversight can promote beneficial deployment while ensuring aviation safety.

4.6 Artificial Intelligence

Definition and Core Concepts

Artificial intelligence refers to computer systems designed to perform tasks that usually require human intelligence, such as learning from experience, recognizing patterns, making decisions under uncertainty, understanding natural language, and solving complex problems. Modern AI is primarily driven by machine learning (ML), in which algorithms learn patterns from data.

Deep learning has driven significant progress in computer vision, natural language processing, speech recognition, and strategic game playing (LeCun, Bengio, & Hinton, 2015). Recently, generative AI has broadened AI's capabilities—large language models (LLMs) and image generation systems now produce new text, code, audio, and visual content. These foundation models trained on large-scale data offer substantial opportunities but also pose risks, including bias amplification, misinformation, and environmental impacts from intensive computational training (Bommasani et al., 2021; Russell & Norvig, 2020).

Global State of the Art

Global investment in AI has surged, reaching about $196 billion in 2023, with estimates surpassing $500 billion by 2027 (Stanford HAI, 2024). However, the worldwide AI landscape is still largely controlled by a few leading tech companies in the United States and China. AI applications now span nearly every sector of the global economy—healthcare diagnostics, financial fraud detection, transportation, manufacturing, and creative industries. The European Union's AI Act (2024) establishes the world's first comprehensive, risk-based regulatory framework for AI systems. The AU Continental AI Strategy (2024) supports

AI adoption across key sectors such as agriculture, health, education, and governance, while simultaneously building foundational capabilities in data infrastructure, computing capacity, skills, and research ecosystems. It emphasizes ethical and responsible AI through governance frameworks that address risks and mobilization of public and private investment to grow AI innovation and markets.

Africa's Current Adoption and Use Cases

Africa's AI ecosystem, though still growing, is expanding rapidly. By 2024, the continent had more than 570 AI-focused startups, with South Africa (145), Nigeria (112), Kenya (98), and Egypt (76) serving as regional hubs (Oxford Insights, 2024). Agricultural AI platforms like PlantVillage and Nuru use computer vision to diagnose crop diseases. Healthcare AI includes Ubenwa (birth asphyxia detection) and radiology interpretation tools. AI-driven credit-scoring platforms such as Jumo, Branch, and Tala have extended credit to over 50 million previously unbanked Africans, promoting financial inclusion. Developing natural language processing for African languages remains a key strategic focus, with initiatives like Masakhane and Lelapa AI.

Development Potential

AI holds significant potential to transform various aspects of African development. In agriculture, AI-powered advisory systems have increased yields by 20–30% in controlled field trials. In healthcare, AI diagnostic tools can provide specialist expertise to remote clinics. Financial inclusion through AI-based credit scoring has facilitated over 50 million first-time formal loans. In education, adaptive learning platforms have shown 15–25% improvements in learning outcomes. Vinuesa et al. (2020) found that AI could positively impact 134 of 169 SDG targets (79%) but may hinder 59 targets (35%) without safeguards.

Figure 4.4. Artificial Intelligence in Africa: Investment Trends (left) and Startup Distribution by Country (right)

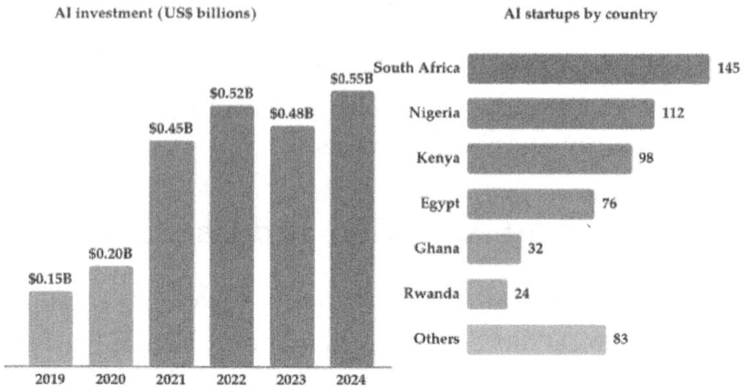

Sources: Based on Stanford (2024); GSMA (2023)

Risks and Challenges

AI deployment in Africa faces significant structural risks. Data scarcity and bias remain major challenges—medical datasets often underrepresent African populations, and facial recognition systems have shown error rates up to 35% higher for dark-skinned individuals (Buolamwini & Gebru, 2018). Infrastructure challenges concentrate benefits in urban centers. Broader political economy risks include concerns about technological sovereignty and digital colonialism.

4.7 Blockchain and Distributed Ledger Technologies

Definition and Core Concepts

Blockchain is a distributed ledger technology (DLT) that records transactions across a decentralized network of computers, ensuring immutability, transparency, and cryptographic security without depending on centralized intermediaries. Public blockchains like Bitcoin and Ethereum are permissionless and open to everyone, while private blockchains limit access to authorized entities. NFTs, DeFi, and CBDCs represent different but increasingly interconnected pillars of the evolving digital asset landscape built on blockchain technology.

Decentralized Finance (DeFi) provides the infrastructure for open, decentralized financial services such as lending, borrowing, and trading. Non-Fungible Tokens (NFTs) are unique assets that represent ownership of digital art, music, or virtual real estate, while Central Bank Digital Currencies (CBDCs) are centralized, regulated digital currencies issued by governments. Unlike cryptocurrencies (e.g., Bitcoin), which are fungible and interchangeable, each NFT is unique and cannot be replaced. NFTs and DeFi generally operate permissionlessly on public blockchains, allowing anyone to create, trade, or interact without needing approval from a central authority. In contrast, CBDCs are typically permissioned, controlled by central banks, and require authorized access (Tapscott & Tapscott, 2016; Zetzsche et al., 2020).

Global State of the Art

The global blockchain market was valued at around $17 billion in 2023 and is projected to exceed $94 billion by 2027 (MarketsandMarkets, 2024). Enterprise adoption has shifted from cryptocurrency speculation to applications in supply chain management, financial services, healthcare records, and identity verification. More than 130 countries are exploring or testing central bank digital currencies (CBDCs) to improve payment efficiency, security, and traceability (Auer, Cornelli, & Frost, 2020). The EU's Markets in Crypto-Assets (MiCA) regulation provides the first comprehensive regional framework for regulating crypto-assets.

Africa's Current Adoption and Use Cases

Economic constraints and governance issues influence Africa's adoption of blockchain technology. Sub-Saharan Africa receives approximately $48 billion in remittances each year, with transaction fees averaging 8–9 percent (World Bank, 2024). Blockchain-based payment platforms have emerged as alternatives.

Blockchain is increasingly used for product traceability. Bext360's blockchain platform has been used to track coffee from Ethiopian and Rwandan farms to export markets. Digital identity solutions are being tested for refugees and informal populations. Land registry projects in Ghana and Kenya aim to improve transparency. Nigeria launched the eNaira in 2021 as Africa's first CBDC (Central Bank of Nigeria, 2021). South Africa's Project Khokha tested wholesale CBDC applications.

Figure 4.5. Blockchain and DLT Applications in Africa: Current Adoption Level vs. Development Impact Potential

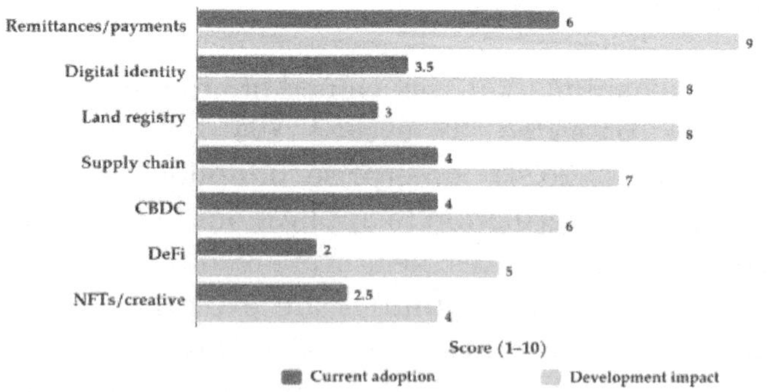

Source: Author's Analysis

Development Potential and Risks

Blockchain technology has strong growth potential for Africa in financial services, land governance, agriculture, energy, and climate finance. Blockchain-based payment systems can lower the high costs of cross-border remittances. Under AfCFTA, distributed ledger technology could improve customs documentation and trade finance. However, issues such as energy efficiency (especially for proof-of-work blockchains), scalability challenges, regulatory uncertainty, cryptocurrency volatility, and low digital literacy present significant challenges. Sustainable blockchain deployment depends on clear regulatory frameworks, regional harmonization, and institutional capacity.

4.8 Big Data Analytics

Definition and Core Concepts

Big data analytics uses tools, architectures, and computational methods to process, analyze, and derive actionable insights from large, complex datasets. The five Vs define big data: volume, velocity, variety, veracity, and value. V**olume** refers to the vast amounts of data generated; **velocity**, the speed at which data is produced and processed; **variety**, the multiple formats and sources of data (both structured and unstructured); **veracity**, the accuracy and reliability of data; and **value**, the ability to extract meaningful insights and benefits from the data. Advanced analytics combines statistical modeling, machine learning, and visualization tools to produce predictive and prescriptive insights (Mayer-Schönberger & Cukier, 2013; Kitchin, 2014).

Global State of the Art

The global big data analytics market was valued at around $274 billion in 2023 and is projected to exceed $650 billion by 2029

(Mordor Intelligence, 2024). Uses include healthcare, finance, agriculture, urban planning, climate science, and government.

Africa's Current Adoption and Use Cases

African big data applications are growing, using mobile phone data, satellite imagery, and digital transaction records. Public health surveillance was transformed during the COVID-19 pandemic. Financial services use transaction analytics to determine credit scores and detect fraud. Agricultural forecasting integrates satellite images, weather data, and soil data. Urban planning in Kigali, Nairobi, Lagos, and Cape Town depends on transport, utility, and demographic data analytics.

Figure 4.6. Big Data Analytics Applications in Africa: Current Adoption vs. Development Impact Potential

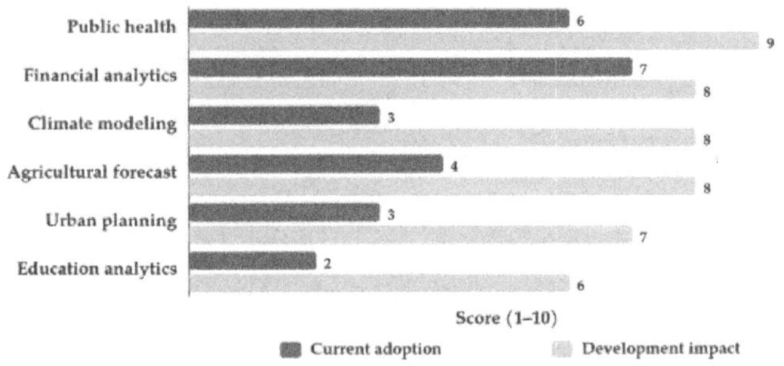

Source: Author's Analysis.

Development Potential and Risks

Big data analytics holds transformative potential by supporting evidence-based decisions across various sectors. Challenges include issues with data quality and completeness, limited storage and processing capacity, governance concerns balancing open data with privacy and sovereignty, skills shortages, and high platform costs. The digital divide means underrepresented groups face systematic blind spots in datasets.

4.9 Internet of Things (IoT)

Definition and Core Concepts

The Internet of Things (IoT) refers to networks of connected physical devices equipped with sensors, software, and communication tools that enable them to collect, exchange, and act on data without continuous human oversight. IoT architecture is typically divided into four functional layers: sensing, connectivity, platform, and application (Atzori, Iera, & Morabito, 2010).

Global State of the Art

The global IoT market exceeded $800 billion in 2023, with over 16 billion connected devices worldwide (IoT Analytics, 2024). Industrial IoT applications dominate in manufacturing, energy, and logistics. Smart city IoT solutions span sectors such as transportation, utilities, waste management, and environmental monitoring.

Africa's Current Adoption and Use Cases

Africa's IoT ecosystem has rapidly expanded, with connected devices rising from about 85 million in 2019 to an estimated 280 million in 2024 (GSMA, 2023). Smart agriculture dominates the sector—soil moisture sensors, livestock tracking systems, and hyperlocal weather stations. Beyond agriculture, IoT deployments improve smart grids, water management, logistics, and healthcare.

Figure 4.7. Internet of Things in Africa: Connected Device Growth (left) and Sectoral Distribution (right)

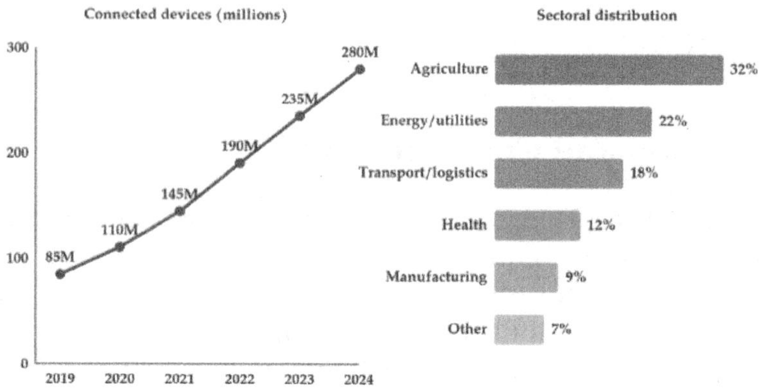

Source: GSMA (2023)

Development Potential and Risks

IoT offers significant growth opportunities by enabling real-time, data-driven management of agriculture, energy, water, transportation, and public health systems. Challenges to IoT deployment include connectivity gaps, limited Low Power Wide Area Network (LPWAN) coverage, energy constraints for powering dispersed sensors, high device costs, interoperability issues, cybersecurity vulnerabilities, and data governance concerns.

4.10 Robotics

Definition and Core Concepts

Robotics is a multidisciplinary field that combines mechanical engineering, electronics, computer science, and artificial intelligence. Robots integrate sensing, actuation, control systems, and computational intelligence. Industrial robots are fixed-position machines used in manufacturing; service robots operate in less-structured environments; and collaborative robots (cobots) work safely alongside humans (Siciliano & Khatib, 2016).

Global State of the Art

The global robotics market was valued at about $55 billion in 2023, with over 500,000 industrial robots installed annually (International Federation of Robotics, 2023). China, Japan, South Korea, Germany, and the United States lead in adoption. Service robotics is rapidly expanding in logistics, healthcare, and agriculture.

Africa's Current Adoption and Use Cases

Robotics adoption in Africa remains limited due to low labor costs, a small manufacturing base, high capital costs, and skills shortages. South Africa leads with about 50 robots per 10,000 manufacturing workers (compared to 397 in South Korea). Egypt's textile and manufacturing sectors have only modest automation. Educational robotics programs are expanding across Africa.

Figure 4.8. Robotics: Global vs. Africa

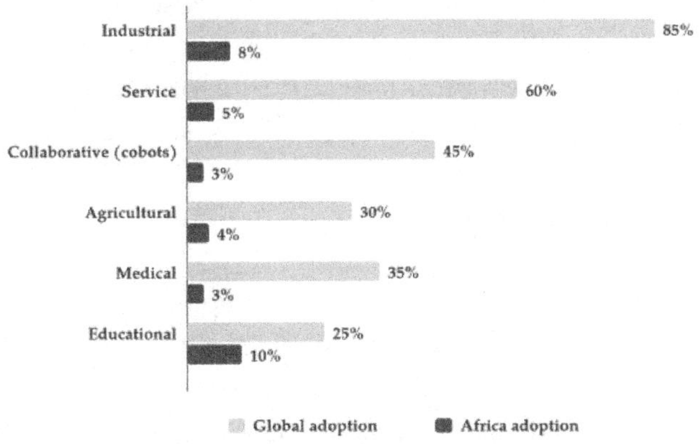

Source: Author's Analysis

Development Potential and Risks

Robot's growth potential depends on specific applications: agricultural robots, healthcare robots for surgical support, educational robots for STEM engagement, service robots for logistics, and collaborative robots (cobots) in manufacturing. The main concern is job displacement in a continent already facing high youth unemployment (International Labour Organization, 2023). Other challenges include high capital costs, maintenance requirements, energy consumption, and ethical issues. Open-source platforms like the Robot Operating System (ROS) and declining component costs create opportunities for African-designed robots.

4.11 Distributed and Renewable Power Systems

Definition and Key Concepts

Distributed and renewable power systems include technologies for generating, storing, managing, and distributing electricity from renewable sources at different scales—from solar home systems serving single households to large solar and wind farms producing hundreds of megawatts. These technologies comprise solar photovoltaics, wind turbines, small hydropower, geothermal energy, modern biomass, battery storage, mini-grids and microgrids, and smart grid systems (IRENA, 2022).

Global State of the Art

Solar PV module costs have fallen by over 90 percent since 2010, and wind turbine costs have declined by about 70 percent. In 2023, renewables represented more than 30 percent of global electricity generation (IEA, 2023). The EU's Green Deal targets 42.5 percent renewable energy by 2030. China installed more solar capacity in 2023 than the rest of the world did in 2022.

Battery storage costs have dropped 89 percent since 2010. Green hydrogen is emerging as a promising solution for sectors that are difficult to decarbonize.

Africa's Current Adoption and Use Cases

Africa's renewable energy revolution is arguably the most advanced example of technology adoption on the continent. Solar home systems with pay-as-you-go (PAYG) financing—pioneered by M-KOPA, BBOXX, d.light, and others—have connected over 8 million households (GOGLA, 2024).

There are more than 3,000 mini-grids across Africa. Utility-scale projects include Morocco's Noor-Ouarzazate solar complex (580 MW), Egypt's Benban Solar Park (1,650 MW), Kenya's Lake Turkana Wind Power (310 MW), South Africa's REIPPPP program (over 6,000 MW), and Ethiopia's Grand Ethiopian Renaissance Dam (6,000 MW). Green hydrogen opportunities include Namibia's Hyphen project and South Africa's Hydrogen Society Roadmap.

Figure 4.9. Distributed and Renewable Power Systems: Installed Capacity Growth (left) and Renewable Mix (right) in Africa

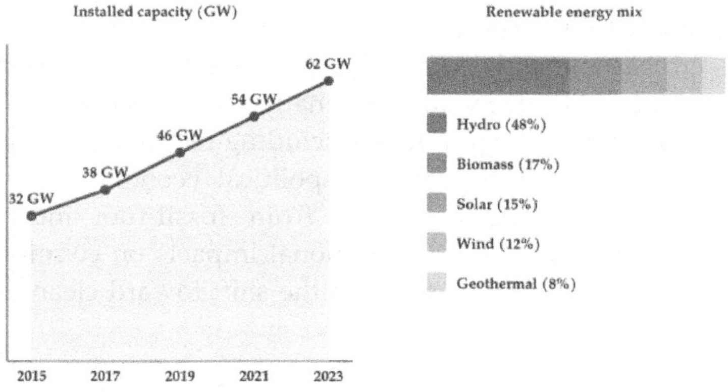

Sources: IRENA (2022); IEA (2023)

Development Potential and Risks

Africa possesses vast renewable energy potential—10 TW of solar, 350 GW of hydropower, 110 GW of wind, and 15 GW of geothermal—far exceeding projected demand (IRENA, 2022). Opportunities include universal energy access, a competitive advantage in energy-intensive industries, leadership in climate efforts, local manufacturing, and energy exports.

Challenges to widespread use of renewable energy include grid integration, financing gaps, policy and regulatory uncertainty, dependence on supply chains, end-of-life management, and the political economy of energy transitions. Integrating variable sources such as solar and wind into existing electricity systems requires stronger transmission networks, energy storage, and smart-grid management to maintain reliability.

Financing gaps also remain significant, especially in Africa, where high upfront investment costs, limited access to long-term capital, and perceived risks discourage investors. Policy and regulatory uncertainty—such as inconsistent incentives, unclear tariffs, and weak regulatory frameworks—can further slow deployment.

In addition, renewable technologies often depend on global supply chains for components such as solar panels, batteries, and critical minerals, creating vulnerabilities to geopolitical disruptions and price volatility. Finally, issues related to end-of-life management of equipment (including recycling solar panels and batteries) and the broader political economy of energy transitions—such as resistance from fossil-fuel industries, employment shifts, and distributional impacts on communities—can complicate and slow down the shift toward clean energy systems.

4.12 Interconnection, Potentials and Risks of Emerging Technologies

The emerging technologies discussed in this chapter are not isolated tools but tightly interconnected systems that enhance each other's impact across the continent. Artificial intelligence depends on data streams created by IoT sensors installed on farms, water systems, and energy grids. At the same time, blockchain provides the trust and transparency layers that make AI-driven financial services and supply chains viable.

Mobile platforms serve as the delivery channels through which millions of Africans access these capabilities, and renewable energy provides the power without which none of these digital systems can operate. Any strategy that treats these technologies in silos—pursuing an AI initiative here, a blockchain project there—will likely perform poorly because real development benefits come from the intersections where solar-powered IoT sensors send data to AI models, whose results are recorded on blockchain ledgers and accessed via mobile money platforms.

Africa's role in this technological landscape is promising, despite limited governance capabilities and weaker foundations. In drone logistics, mobile banking, pay-as-you-go solar power, and digitally connected agriculture, African innovators have created solutions that are not just imported but are genuinely top-tier worldwide, driven by the need to solve urgent problems with limited infrastructure. This track record shows that when the right conditions—such as enough energy, connectivity, skills, and supportive institutions—are in place, African enterprises and institutions can skip traditional development steps and set global standards rather than follow them.

The potential of these ten frontier technologies comes with risks that require equal integrated attention. Data bias can worsen inequalities that AI aims to address, skills gaps and brain drain threaten to weaken the human capital needed for local innovation, and the lack of strong governance frameworks leaves societies exposed to harms ranging from financial fraud to environmental damage.

4.13 Conclusion

This chapter provides systematic evaluations of 10 emerging technology domains. Several cross-cutting findings emerge.

- First, Africa is not just a passive recipient of technologies developed elsewhere—in drone delivery, mobile financial services, off-grid renewable energy, and agricultural platforms, African innovators have developed world-leading solutions.

- Second, enabling-layer constraints—energy access, connectivity, data infrastructure, skills, and institutional quality—are consistently present across all technology domains.

- Third, the technologies are highly interconnected—AI relies on data, IoT produces data that AI analyzes, blockchain provides trust infrastructure, and renewable energy powers the entire digital infrastructure.

- Fourth, every technology presents both opportunities and risks, requiring governance frameworks that are balanced, flexible, and rooted in African values.

- Fifth, global comparisons highlight African strategy without taking control of it—lessons from India, the EU, East Asia, and Latin America can be adapted rather than copied.

References

14Trees. (2021). Africa's first 3D-printed house completed in Malawi. 14Trees/Holcim.

African Development Bank. (2023). African economic outlook 2023: Mobilizing private sector financing for climate and green growth in Africa. African Development Bank Group.

Arner, D. W., Barberis, J., & Buckley, R. P. (2017). FinTech and regtech. Journal of Banking Regulation, 19(4), 1–14.

Atzori, L., Iera, A., & Morabito, G. (2010). The Internet of Things: A survey. Computer Networks, 54(15), 2787–2805.

AU-NEPAD. (2021). African innovation response to COVID-19: Local manufacturing and 3D printing. African Union Development Agency.

Auer, R., Cornelli, G., & Frost, J. (2020). Rise of the central bank digital currencies. BIS Working Papers No. 880.

Azuma, R. T. (1997). A survey of augmented reality. Presence: Teleoperators & Virtual Environments, 6(4), 355–385.

Bommasani, R., et al. (2021). On the opportunities and risks of foundation models. Stanford HAI.

Brynjolfsson, E., & McAfee, A. (2014). The second machine age. W. W. Norton.

Buolamwini, J., & Gebru, T. (2018). Gender shades. Proceedings of ML Research, 81, 1–15.

Casino, F., Dasaklis, T. K., & Patsakis, C. (2019). Blockchain-based applications. Telematics and Informatics, 36, 55–81.

Central Bank of Nigeria. (2021). Design paper for the eNaira—Central Bank of Nigeria.

Couldry, N., & Mejias, U. A. (2019). The costs of connection: How data is colonizing human life and appropriating it for capitalism. Stanford University Press.

Drone Industry Insights. (2024). Drone market report 2024. Drone Industry Insights.

Gibson, I., Rosen, D., & Stucker, B. (2015). Additive manufacturing technologies (2nd ed.). Springer.

GOGLA. (2024). Global off-grid solar market report 2024. GOGLA.
Grand View Research. (2024). Augmented reality market size, share & trends analysis report. Grand View Research.
GSMA. (2023). The mobile economy in Sub-Saharan Africa 2023. GSMA.
International Energy Agency. (2023). World Energy Outlook 2023. IEA.
International Federation of Robotics. (2023). World robotics report 2023. IFR.
International Labour Organization. (2023). World employment and social outlook 2023. ILO.
International Renewable Energy Agency. (2022). Renewable capacity statistics 2022. IRENA.
IoT Analytics. (2024). State of IoT 2024. IoT Analytics.
Jerald, J. (2015). The VR book: Human-centered design for virtual reality. Morgan & Claypool.
Kitchin, R. (2014). The data revolution. Sage.
LeCun, Y., Bengio, Y., & Hinton, G. (2015). Deep learning. Nature, 521(7553), 436–444.
MarketsandMarkets. (2024). Blockchain market — Global forecast to 2027. MarketsandMarkets.
Mayer-Schönberger, V., & Cukier, K. (2013). Big data. Houghton Mifflin Harcourt.
Mordor Intelligence. (2024). Big data analytics market — Growth, trends, and forecasts. Mordor Intelligence.
Nakamoto, S. (2008). Bitcoin: A peer-to-peer electronic cash system.
Ngo, T. D., Kashani, A., Imbalzano, G., Nguyen, K. T. Q., & Hui, D. (2018). Additive manufacturing (3D printing): A review of materials, methods, applications and challenges. Composites Part B: Engineering, 143, 172–196.
Not Impossible Labs. (2020). Project Daniel: 3D-printed prosthetics for conflict amputees. Not Impossible Labs.
Oxford Insights. (2024). Government AI readiness index 2024. Oxford Insights.

Russell, S., & Norvig, P. (2020). Artificial intelligence (4th ed.). Pearson.
Siciliano, B., & Khatib, O. (2016). Springer handbook of robotics (2nd ed.). Springer.
Stanford HAI. (2024). AI Index Report 2024. Stanford University Human-Centered AI Institute.
Statista. (2024). Virtual reality (VR) — Statistics & facts. Statista.
Tapscott, D., & Tapscott, A. (2016). Blockchain revolution. Penguin.
Valavanis, K. P., & Vachtsevanos, G. J. (2015). Handbook of Unmanned Aerial Vehicles. Springer.
Vinuesa, R., et al. (2020). The role of AI in achieving the SDGs. Nature Communications, 11, 233.
Wohlers Associates. (2023). Wohlers Report 2023: 3D printing and additive manufacturing global state of the industry. Wohlers Associates.
World Bank. (2024). World development indicators 2024. World Bank Publications.
Zetzsche, D. A., et al. (2020). Decentralized finance. Journal of Financial Regulation, 6(2), 172–203.
Zipline. (2024). Impact report: 10 million deliveries. Zipline International.

PART III

Emerging Technologies and African Socio-economic Development

Chapter 5: Aligning Emerging Technologies with Africa's Development

5.1 Introduction

Africa's engagement with emerging technologies must align with the continent's aspirations. The continent has a clearly defined long-term development vision through AU Agenda 2063: The Africa We Want—that was adopted by African Union Heads of State and Government in 2013 and brought to life through successive ten-year implementation plans. This vision, spanning fifty years from 2013 to 2063, provides the strategic framework for assessing the potential of emerging technologies, prioritizing opportunities, and managing risks. Technologies that support Agenda 2063's goals should be accelerated; those that could undermine them should be subject to governance safeguards.

This chapter connects the technical analysis of frontier technologies in Part II with the sector-specific applications discussed in Part III, creating a strategic link between technological capabilities and Africa's development goals. It begins with an overview of Agenda 2063, reviewing progress and highlighting where technology can help address implementation gaps. Next, it links specific technologies to each of the seven aspirations, illustrating how artificial intelligence, blockchain, the Internet of Things, drones, renewable energy systems, and other emerging technologies can foster continental transformation. The chapter also examines Agenda 2063's flagship programs, focusing on the technological requirements of initiatives such as the African Continental Free Trade Area and the Programme for Infrastructure Development in Africa. Finally, it discusses how regional integration and continental public goods—such as shared infrastructure, computing facilities, data governance frameworks, and technology standards—can enhance the collective benefits of technology adoption across the African Union's 55 Member States.

5.2 Overview of African Union Agenda 2063

The Seven Aspirations of "The Africa We Want"

Agenda 2063 outlines seven interconnected aspirations that collectively define the continent's development vision. These aspirations are more than mere statements; they form the basis for planning at the continental, regional, and national levels. Recognizing the technological dimensions of each aspiration is essential for effective technology implementation.

- **The first aspiration** envisions a prosperous Africa built on inclusive growth and sustainable development, including industrialization, economic diversification, modernized agriculture, and a well-connected continent with world-class infrastructure.

- **The second aspiration** envisions a united continent, politically connected through Pan-African ideals, with world-class infrastructure, including high-speed rail systems, roads, shipping routes, and a strong ICT and digital economy.

- **The third aspiration** envisions good governance, democracy, respect for human rights, justice, and the rule of law, all of which require transparent and accountable institutions.

- **The fourth aspiration** seeks a peaceful and secure Africa, addressing conflict prevention, resolution, and management.

- **The fifth aspiration** encourages a robust cultural identity, shared heritage, common values, and ethics.

- **The sixth aspiration** pursues an Africa where development is driven by its people, especially women and youth.

- **The seventh aspiration** positions Africa as a strong, united, resilient, and influential global player and partner (AU, 2015).

Progress Assessment and Implementation Gaps

As Africa entered the 'Decade of Acceleration' (2024–2033) under the Second Ten-Year Implementation Plan, the continent has reached important milestones, though significant gaps still exist. The African Continental Free Trade Area, launched in January 2021, and its Digital Trade Protocol, adopted in February 2024, stand out as the most notable institutional achievements toward continental integration.

The African Passport and Free Movement regime has progressed with visa liberalization reforms—Ghana's 2025 visa-free announcement marking a historic milestone—while the Single African Air Transport Market has increased its membership. Monitoring and accountability have improved through biennial reviews and public dashboards that track implementation (Africa Center, 2025).

However, significant implementation gaps still exist. By 2025, only about 40% of Africa's population had broadband Internet access, which limits participation in the adoption of emerging technologies. Integration into global value chains remains below 15% of GDP. Infrastructure gaps remain in connectivity, energy, and transportation.

The AU Digital Transformation Strategy for Africa (2020–2030) was elevated to a flagship project of Agenda 2063 at the 2024 Summit, and the African Digital Compact, adopted in 2024, provides an additional framework. However, turning these frameworks into actual actions requires technological capabilities that many AU Member States currently lack. These gaps are exactly where emerging technologies have the greatest potential to transform—not as replacements for institutional development but as catalysts that can reduce timelines and enable leapfrogging of traditional development stages.

Technology's Transformative Role in Agenda 2063

Technology plays a complex role in Agenda 2063's framework. It serves both as an enabler for achieving goals across all seven aspirations and as a source of disruption that could worsen inequalities if not properly managed. The Agenda's vision of Africa as a 'global powerhouse of the future' implicitly depends on technological capabilities. However, the Agenda was developed before the recent rise of generative AI, advanced blockchain applications, IoT expansion, big data analytics, and drone logistics—technologies that greatly expand what is possible within the Agenda's timeframe.

The key strategic question is not whether Africa should adopt emerging technologies, but how to align adoption with the continent's self-defined development goals instead of external technological agendas. This involves mapping specific technologies to particular aspirations, identifying where technological capabilities provide the greatest development benefits, and establishing governance frameworks that ensure technology supports the Agenda's inclusive, people-centered development vision.

5.3 Mapping Technologies to Agenda 2063 Aspirations

This section systematically links emerging technologies to each of Agenda 2063's seven aspirations, highlighting specific applications, African examples, and implementation needs. The mapping is not merely illustrative; it aims to provide stakeholders with a practical framework for technology prioritization.

Aspiration 1: A Prosperous Africa – Inclusive Growth and Industrialization

The first aspiration's focus on inclusive growth, industrialization, economic diversification, and modern agriculture opens up significant opportunities for emerging technologies. AI and automation are transforming manufacturing—Egypt's textile industry and South Africa's automotive sector are already adopting Industry 4.0 technologies, such as digital twins, predictive maintenance, and robotic assembly.

AI could add an estimated USD 2.9 trillion to Africa's economy by 2030, representing roughly a three-percentage-point increase in annual GDP growth (GSMA, 2024; Mastercard, 2025). By mid-2025, African AI startups had raised USD 803.2 million, with applications across healthcare, agriculture, logistics, and natural language processing in African languages.

Fintech has become Africa's most developed technology sector, with over 1,000 firms expected by 2024 and several reaching unicorn valuations—Flutterwave at USD 3 billion, OPay at USD 2 billion, and Moniepoint, which achieved unicorn status in October 2024. Mobile money's 1.1 billion registered accounts by 2024 will lay the foundation for broader digital financial inclusion. Africa's fintech revenue is projected to grow 13-fold to USD 65 billion by 2030 (Global FinTech; DevelopmentAid, 2025). Big data analytics supports economic planning, with central banks in the COMESA region already using AI and big data to improve economic forecasting and banking oversight.

Agricultural technology focuses on the sector that employs more than half of Africa's workforce. The adoption of emerging technologies is increasing yields through AI-driven pest detection, satellite-based crop monitoring, blockchain-supported supply chain traceability, and IoT-enabled precision irrigation. Combining these technologies across the agricultural value chain — from precise inputs and smart monitoring to post-harvest management and digital market access — produces significant productivity gains that traditional methods cannot match.

Figure 5.1. AI Impact Potential by Sector

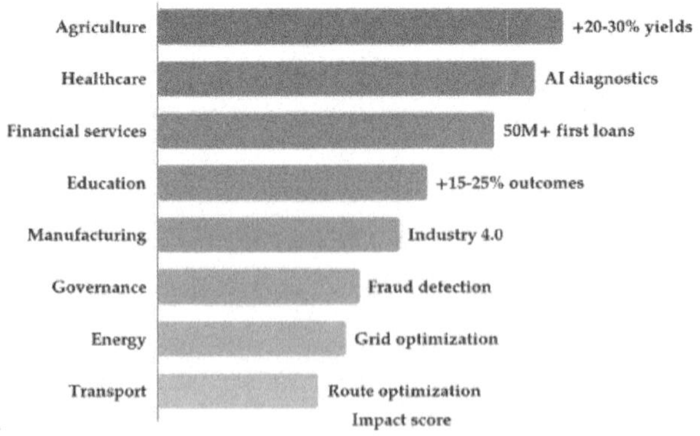

Source: GSMA (2024); Mastercard (2025); Partech Africa (2024)

Aspiration 2: An Integrated Continent – Infrastructure and Connectivity

Continental integration requires infrastructure that connects the African Union's 55 countries physically, digitally, and commercially. The Internet of Things supports smart infrastructure management—sensors that monitor road conditions, bridge integrity, and port operations across the continent. Digital payment systems, centered on the Pan-African Payment and Settlement System (PAPSS), facilitate trade integration by enabling instant cross-border transactions in local currencies. PAPSS has expanded to over 13 West African states, as well as Kenya, Malawi, Rwanda, and Zambia, and launched PAPSSCARD in June 2025 as the continent's first pan-African card scheme.

Blockchain technology improves trade by enabling digital certificates of origin, smart contracts, and supply chain traceability, thereby building the trust required to increase intra-African commerce. The AfCFTA Digital Trade Protocol, with its eight annexes covering digital payments, data transfers, digital identity, and emerging technologies, provides the legal framework for technology-driven integration.

Investments in subsea cables, along with terrestrial fiber expansion, are expected to triple Africa's data center market to over USD 3 billion by 2030 (WEF, 2025). However, much work remains to develop backbone networks across cities and borders to harness emerging technologies for an integrated continent.

Aspiration 3: Good Governance – Democracy and the Rule of Law

Technology provides powerful tools to improve governance, transparency, and accountability, but it also introduces risks of surveillance and control that need careful management.

Blockchain systems increase transparency in areas such as public procurement, land registries, and elections. Ghana's blockchain land registry pilot shows how distributed ledger technology can secure property rights, reduce disputes, and build citizen trust in government records. Digital identity systems—when designed with privacy protections and including marginalized groups—enable citizens to access government services, participate in democratic processes, and exercise economic rights.

E-governance platforms in countries such as Rwanda, Kenya, Mauritius, Morocco, South Africa, and Zambia digitize government services, including business registration and tax filing, helping reduce opportunities for corruption stemming from discretionary gatekeeping. AI-powered analytics detect anomalies in public spending, identify fraudulent procurement patterns, and enhance customs enforcement.

However, these technologies also pose governance risks: AI surveillance systems can enable authoritarian monitoring; digital identity requirements might exclude undocumented populations; and e-governance platforms can centralize authority, weakening democratic accountability. Aspiration 3 stresses that technology used in governance must be democratically overseen, incorporate human rights safeguards, and adhere to principles of inclusive design.

Aspiration 4: A Peaceful and Secure Africa

Conflict prevention, resolution, and management increasingly involve technology. Early warning systems driven by AI and big data analytics can identify escalation patterns, forecast displacement, and guide preventive actions. The African Union's Continental Early Warning System uses satellite imagery, social media monitoring, and machine-learning algorithms to detect signs of conflict faster than traditional intelligence methods.

Cybersecurity technologies protect critical infrastructure—such as financial systems, energy grids, and telecommunications networks—from attacks that could destabilize economies and societies. Continental cybersecurity collaboration through Computer Emergency Response Team (CERT) networks promotes collective defense against threats that cross national borders.

Border management technologies, including biometric identification, drone surveillance, and IoT-based monitoring, can enhance security while—if properly managed—supporting the free movement of people envisioned by the African Passport initiative. However, the dual-use nature of these security tools requires strong governance: surveillance methods used for border security can be misused against political opponents; facial recognition systems carry risks of ethnic bias; and militarized drone technologies raise concerns under humanitarian law. Aspiration 4's peace agenda mandates that the deployment of security technology must adhere to human rights principles and be subject to independent oversight.

Aspiration 5: A Strong Cultural Identity

Africa's cultural and linguistic diversity—with over 2,000 languages spoken across the continent—poses both a technological challenge and a unique opportunity. Augmented and virtual reality technologies help preserve and promote cultural heritage: virtual museum tours showcase Africa's archaeological treasures to a global audience; AR overlays bring historical sites to life for visitors and students; and immersive VR experiences protect intangible cultural heritage, including oral traditions, musical performances, and ceremonial practices.

AI-powered natural language processing for African languages is arguably the most culturally significant application of technology. Initiatives such as Masakhane (a grassroots natural language processing research community), Lelapa AI, and language-specific models for Swahili, Yoruba, Twi, Xhosa, Zulu, and other African languages are developing the computational tools these languages need for communities to take part in the AI revolution. These efforts go beyond translation; they enable search engines, voice assistants, educational content, and government services to be accessible in the languages that communities actually speak, helping preserve linguistic heritage and promote digital inclusion. Digital archives, blockchain-verified provenance for cultural artifacts, and digital rights management for African creative content support these initiatives in fostering a technology-powered cultural renaissance.

Aspiration 6: People-Driven Development

The sixth aspiration's focus on human development—especially women and youth empowerment—aligns with technologies that expand access to education, healthcare, and climate adaptation tools. EdTech platforms, such as Eneza Education and Ubongo, support students' learning through mobile-first approaches suited to low-bandwidth environments. AI-driven adaptive learning systems personalize educational experiences, while VR simulations provide virtual laboratory access for STEM education in resource-limited settings.

Digital health technologies tackle Africa's urgent healthcare workforce shortages. AI-powered diagnostics for malaria, tuberculosis, and cervical cancer offer exceptional accuracy, enabling faster diagnosis and earlier treatment. Telemedicine platforms connect patients in remote areas with specialists in urban centers and abroad. Climate adaptation technologies—satellite-based early warning systems, AI-driven weather forecasting, IoT-enabled precision irrigation—assist communities in building resilience against climate shocks that mainly affect Africa's most vulnerable populations.

Aspiration 7: Africa as a Strong Global Partner

The seventh aspiration positions Africa not as a passive recipient of global technology but as an active contributor to technological innovation and international governance. Africa's mobile money revolution has already demonstrated the continent's ability to pioneer technologies that are later adopted worldwide. The continent's strategies for drone regulation (such as Rwanda's and Ghana's frameworks), fintech sandboxes, and mobile-first digital services are offering governance innovations that other regions are increasingly researching and adopting.

Digital sovereignty and data governance are crucial to Africa's global influence. The African Union's Data Policy Framework, the African Digital Compact adopted in 2024, and the Malabo Convention on Cybersecurity and Personal Data Protection establish continental structures that give Africa a voice in shaping global digital rules.

South-South technology cooperation—with partners in Asia, Latin America, and the Middle East—offers alternative methods for technology transfer that complement traditional North-South approaches. Africa's rich deposits of cobalt, lithium, graphite, and rare-earth minerals—critical for digital and green sectors—provide it with strategic leverage in negotiating technology deals that include knowledge sharing and local value creation, not just raw material extraction.

Table 5.1. Technology–Aspiration Alignment: Key Applications by Domain

Technology Domain	Prosperity	Integration	Governance	Peace & Security	People-Driven Dev.
AI & Machine Learning	Smart agriculture, Industry 4.0, credit scoring	Trade analytics, corridor optimization	Fraud detection, spending analysis	Early warning, conflict prediction	Adaptive learning, health diagnostics
Blockchain & DLT	Supply chain traceability, fintech	Digital certificates of origin, PAPSS	Land registries, procurement transparency	Secure identity, border management	Health records, credential verification
Internet of Things	Precision irrigation, smart manufacturing	Infrastructure monitoring, smart ports	Service delivery tracking	Border monitoring, critical infrastructure	Remote health monitoring, climate sensors
Drones & Robotics	Agricultural spraying, logistics	Infrastructure inspection	Service delivery to remote areas	Surveillance, humanitarian aid	Medical delivery, educational access
Renewable Energy Tech	Off-grid energy, pay-as-you-go solar	Regional power pools, grid interconnection	Energy access transparency	Energy security for critical systems	Community energy, climate resilience
AR/VR & Digital Media	Skills training, design prototyping	Virtual trade fairs, remote collaboration	Citizen engagement platforms	Conflict simulation, training	Cultural preservation, immersive education

Source: Author's Analysis

5.4 Flagship Programmes and Technology Enablement

Agenda 2063 includes specific flagship programs aimed at driving socio-economic transformation. Each program has unique technology requirements that influence the pace and success of implementation. This section explores how emerging technologies can accelerate the delivery of the most technology-driven flagship initiatives.

African Continental Free Trade Area (AfCFTA)

The AfCFTA is the most advanced flagship program of Agenda 2063, with its Digital Trade Protocol adopted in February 2024 and eight supplementary annexes approved in February 2025. Technology enablement for AfCFTA covers several areas, including digital trade infrastructure such as electronic customs systems, digital certificates of origin, and paperless trade platforms; blockchain-based supply chain traceability for rules of origin verification; AI-powered trade analytics to identify market opportunities and optimize trade corridors; and the PAPSS digital payment infrastructure, which enables seamless cross-border transactions. Full AfCFTA implementation could lift 30 million people out of extreme poverty and increase continental income by USD 450 billion by 2035 (WEF, 2025)—outcomes that technology deployment can greatly accelerate.

Programme for Infrastructure Development in Africa (PIDA)

The Programme for Infrastructure Development in Africa (PIDA) was launched in 2012 by the African Union Commission, the African Union Development Agency (AUDA–NEPAD), the African Development Bank, and the United Nations Economic Commission for Africa to address Africa's persistent infrastructure gap and improve regional integration. PIDA seeks to support Africa's development objectives by encouraging cross-border infrastructure investments that connect markets, reduce trade costs, and increase access to transportation, energy, digital services, and water resources across the continent (African Union Commission, African Development Bank, & NEPAD, 2012).

Through its Priority Action Plans (PAPs), the program creates strategic projects that can boost regional connectivity and open up economic opportunities. Initial PIDA efforts centered on regional transport corridors, power pools, and fiber-optic backbone networks linking African countries and regions. PIDA Priority Action Plan 2 (PIDA-PAP 2) for 2021–2030, which extends the list of priority projects from PAP 1 (2010-2020), highlights climate resilience, green energy, and digital infrastructure. The digital portfolio focuses on developing regional fiber networks and digital platforms.

Emerging technologies can greatly improve the planning, implementation, operation, and impact of infrastructure projects under PIDA. Since PIDA emphasizes continental infrastructure in transport, energy, ICT, and water systems, technologies such as AI, IoT, blockchain, robotics, additive manufacturing, and renewable energy can, over time, boost project efficiency, transparency, resilience, and sustainability. The relationship is complementary: PIDA provides the physical infrastructure backbone for Africa, while emerging technologies enhance its design, management, and utilization.

African Passport and Free Movement of People

The African Passport initiative requires a digital identity system that facilitates seamless cross-border verification while protecting privacy and preventing misuse. Biometric identification, blockchain-verified credentials, and interoperable digital identity platforms form the technical foundation for the

African Passport. Ghana's 2025 visa-free move and Ethiopia's and Rwanda's visa liberalization demonstrate political commitment; technology implementation makes this a reality through automated border control, digital visa processing, and trusted traveler programs that support movement while maintaining security.

Silencing the Guns

The 'Silencing the Guns by 2030' initiative—extended due to ongoing conflicts—relies on technology-driven conflict prevention and management. The Continental Early Warning System uses AI and satellite imagery to predict conflicts. Cybersecurity measures protect vital infrastructure from destabilizing attacks. Communication technologies facilitate mediation and peace efforts over long distances. However, the dual-use nature of security technologies means the 'Silencing the Guns' initiative must explicitly address the governance of military AI, concerns about autonomous weapons, and the prevention of technology-enabled repression.

Education, Skills, and the Demographic Dividend

The AU's designation of 2025–2034 as the 'Decade of Education and Skills Development' establishes an institutional mandate for technology-enabled education across the continent. AI-powered learning platforms, coding bootcamps, online apprenticeships, micro-credential platforms, and STEM promotion through educational robotics all help build the human capital needed to translate demographic potential into economic reality.

5.5 Regional Integration and Continental Public Goods

Technology adoption reaches its full potential when supported by continental public goods—shared infrastructure, harmonized standards, and collaborative frameworks that no single country can efficiently provide alone. This section explores four key domains of continental public goods vital for technology-driven development in Africa.

Pan-African Digital Infrastructure

Continent-wide digital infrastructure is a crucial public good that enables all technology applications. This includes shared submarine cable landing stations and land-based backbone networks connecting landlocked countries; Internet Exchange Points that keep intra-African traffic local instead of routing through Europe or North America; pan-African cloud infrastructure and data centers hosting African content and applications within the continent; and spectrum harmonization that allows economies of scale in mobile network deployment and device manufacturing.

Shared Research and Innovation Facilities

Africa's research infrastructure, such as compute resources, remains significantly underdeveloped relative to the continent's needs and potential. Shared facilities—such as compute resources and continental and regional centers of excellence in AI, IoT, renewable energy, and other frontier technologies—can provide the critical mass of equipment, expertise, and collaboration that individual countries cannot sustain.

The African Institute for Mathematical Sciences network, spanning six countries, exemplifies this model: shared investment in centers of excellence that serve continental rather than only national constituencies. Similar models for AI research (building on centers like South Africa's Centre for Artificial Intelligence Research and Kenya's AI labs), genomics (extending the African Genome Variation Project), and biotechnology could greatly expand the continent's research capacity without requiring each country to build duplicative facilities. Regional cooperation is also key in acquiring and sharing compute facilities.

Continental Data Governance Frameworks

Data governance is arguably the most important public good on the continent in the digital age. The AU Data Policy Framework and its implementation plan aim to strengthen and harmonize data governance across Africa, creating a shared data space that balances openness with sovereignty. The African Digital Compact, adopted in 2024, links this effort to Agenda 2063 and the Digital Transformation Strategy. Cross-border data flows are essential for continental platforms in health, agriculture, trade, and climate—yet issues related to data protection, privacy, and sovereignty require governance frameworks that build trust.

The AfCFTA Digital Trade Protocol's annex on cross-border data transfers provides trade-specific governance, while the Malabo Convention establishes broader norms for data protection and cybersecurity. Combining these frameworks into a single continental data governance system—one that enables data flows necessary for technological and development applications while protecting citizens' rights—is an essential public good that no single country can deliver on its own.

Technology Standards Harmonization and Cross-Border Regulatory Cooperation

Fragmented technical standards and regulatory approaches across African countries are a major barrier to deploying technology across the continent. A drone regulatory framework created for Rwanda cannot be used in neighboring Burundi without separate approval. A digital health application approved in Kenya needs separate regulatory processes in each East African Community partner state. Payment technologies licensed in one jurisdiction face new compliance requirements at each border.

Standards harmonization—for IoT devices, AI systems, blockchain applications, drone operations, health technologies, and digital financial services—is a public good that lowers costs, enables scaling, and accelerates adoption. Technology standards harmonisation and cross-border regulatory cooperation are therefore critical. Regional Economic Communities act as natural platforms for gradual harmonization, with successful regional efforts expanding to the continental level.

5.6 Conclusion

The alignment of emerging technologies with Agenda 2063 reveals both significant opportunities and notable governance challenges. All goals—from prosperity to integration, governance, peace, cultural identity, people-centered development, and global partnerships—follow clear technology-driven paths supported by existing African innovations and implementations. The examples in this chapter demonstrate that African-led technological innovation is beginning to contribute to achieving the Agenda's aspirations.

However, realizing the full potential of technology-Agenda 2063 alignment requires addressing four strategic imperatives.

- First, technology deployment must be intentionally aligned with development priorities rather than influenced by technology availability or external agendas—Agenda 2063 and national plans should guide technology choices, not the other way around.

- Second, continental public goods—shared infrastructure, shared compute resources, data governance, standards, and research facilities—are essential for maximizing the continent's collective benefits from technology.

- Third, governance frameworks must address the dual-use nature of many technologies, making sure that tools intended for prosperity, governance, and peace are not misused for repression, exclusion, or growing inequality.

- Fourth, Africa's involvement in global technology governance should position the continent as an active rule-maker—leveraging its demographic strength, resource wealth, and innovation history to set rules rather than accept them.

The 'Decade of Acceleration' (2024–2033) offers both urgency and opportunity. The digital trade infrastructure is being built. National AI strategies are being developed. Innovation ecosystems are growing stronger. The key question is whether these technological capabilities will be used in accordance with Africa's self-defined vision or allowed to develop in fragmented, externally driven, or exclusionary ways. This chapter supports the former, and the following chapters will examine, in sector- and governance-specific detail, how to achieve this intentional alignment in practice.

References

Africa Center for Strategic Studies. (2025, December 8). Assessing progress on Africa's Agenda 2063. https://africacenter.org

African Union. (2015). Agenda 2063: The Africa We Want (Popular Version). AU Commission.

African Union. (2020). The Digital Transformation Strategy for Africa 2020–2030. AU Commission.

African Union. (2024). African Digital Compact. Adopted at the AU Assembly, Addis Ababa.

African Union. (2024). Protocol to the AfCFTA Agreement on Digital Trade—37th AU Assembly.

African Union Commission, African Development Bank, & NEPAD Planning and Coordinating Agency. (2012). Programme for Infrastructure Development in Africa (PIDA), Interconnecting, integrating and transforming a continent. Addis Ababa, African Union.

African Union Development Agency–NEPAD. (2020). PIDA Priority Action Plan (PAP 1) implementation progress report, Midrand: AUDA-NEPAD.

African Union Development Agency–NEPAD. (2021). PIDA Priority Action Plan 2 (2021–2030). Midrand: AUDA-NEPAD.

Bhattacharya, S. (2024, December). Towards Agenda 2063: Africa's digitalisation imperatives. ORF Special Report No. 239. Observer Research Foundation.

DevelopmentAid. (2025, November 18). AI-powered fintech and banking are driving Africa's development. https://www.developmentaid.org

GSMA. (2024, July). AI for Africa: Use cases delivering impact. GSMA Mobile for Development.

Mastercard. (2025, August). AI in Africa: Harnessing the transformative power. Mastercard Centre for Inclusive Growth.

Tech in Africa. (2025, August 5). Crypto, fintech, agritech & AI: Africa's most promising sectors in 2025. https://www.techinafrica.com

World Economic Forum. (2025, July). Africa is harnessing technology to leapfrog towards growth. https://www.weforum.org

Chapter 6: Social and Economic Applications of Emerging Technologies

6.1. Introduction

Emerging technologies are transforming vital social and economic sectors with major implications for human development across Africa and around the world. These technologies have strong potential, shown by digital health platforms that increase healthcare access in remote areas, educational tools that boost learning outcomes, agricultural innovations that improve food security and climate resilience, and smart city solutions that address the problems of rapid urban growth (Sachs et al., 2019; African Union, 2020).

This chapter explores the application of emerging technologies in four key areas—health, education, agriculture, and urban systems—with a primary focus on Africa. To broaden the analysis and provide useful insights, experiences from Asia, Europe, Latin America and the Caribbean and Oceania are included. This approach highlights lessons that can be adapted, potential challenges, and opportunities for collaboration.

The chapter discusses the advantages and disadvantages of these technologies, emphasizing that successful implementation depends not only on the technologies themselves but also on supporting ecosystems that include infrastructure, governance, human capital, and financing. These ecosystems are essential in shaping whether innovation promotes inclusive growth or exacerbates existing inequalities.

6.2 Health Systems and Digital Health

African health systems face significant challenges, including a shortage of healthcare workers (about 2.3 per 1,000 people, compared to the global average of 6.9), long distances between patients and providers, high costs, fragile health information systems, and complex disease burdens involving both infectious and non-communicable diseases (World Health Organization, 2022). Digital health technologies such as telemedicine, AI-powered diagnostics, drone delivery, health data analytics, wearable devices, and robotics offer potential solutions to these structural barriers (Wahl et al., 2018).

The global digital health market was valued at approximately USD 330 billion in 2023 and is projected to exceed USD 650 billion by 2028 (Grand View Research, 2024). Although Africa currently accounts for a small fraction of this market, the continent has become a center for developing contextually relevant health innovations that are now being adopted in other developing regions. Major applications in the health sector driven by emerging technologies include telemedicine and remote care, AI-powered diagnostics, drone delivery, health analytics, and the adoption of IoT and robotics.

Telemedicine and Remote Care

Telemedicine platforms enable remote consultations between patients and healthcare providers using mobile devices or computers. These platforms support a variety of applications, from routine check-ups and triage to specialist referrals, mental health services, and management of chronic diseases. In Africa, telemedicine addresses a major infrastructural gap: the continent's low ratio of 2.3 health workers per 1,000 people, which falls short of the SDG's target of 4.45. This gap requires alternative strategies to expand healthcare access beyond traditional infrastructure development (WHO, 2022). Telemedicine effectively extends the reach of limited healthcare expertise by enabling urban-based specialists to serve geographically dispersed populations.

The COVID-19 pandemic significantly accelerated the adoption of telemedicine in Africa. Consultations rose from approximately 1.2 million in 2019 to 8.5 million in 2020, and reached 22 million by 2024 (Mars, 2020; Africa CDC, 2024). Major platforms contributing to this growth include Babylon Health, which partners with the Government of Rwanda to provide AI-powered symptom assessment and video visits, and mPharma, which offers integrated telemedicine services across 9 African countries. Several countries, including Nigeria, Kenya, South Africa, and Ethiopia, also run national telemedicine programs.

Figure 6.1: Growth of telemedicine consultations in Africa

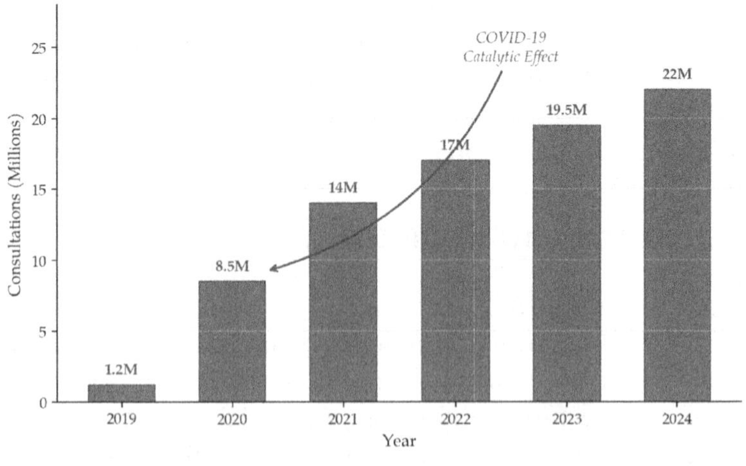

Sources: Mars (2020); Africa CDC (2024)

Despite this progress, significant challenges still impede the wider adoption of telemedicine systems. Connectivity remains a major barrier, especially in rural areas where access is most needed. Many countries lack comprehensive regulatory frameworks that address licensure across jurisdictions, liability for remote diagnoses, and data protection. Furthermore, digital literacy among both patients and healthcare providers limits adoption.

Box 6.1 Comparative Insight: Telemedicine Expansion in Asia and Latin America

India's eSanjeevani platform, launched during COVID-19, facilitated over 100 million teleconsultations by 2023, demonstrating the scalability of government-led telemedicine at a population level (Government of India, 2023).

In Brazil, the Rede Nacional de Telessaude connects more than 35,000 primary care units to specialist centers, significantly reducing unnecessary patient referrals by an estimated 70% (Harzheim et al., 2019). These models provide valuable insights for African health systems regarding public-sector telemedicine infrastructure, the integration of emerging technologies, regulatory frameworks for cross-border practice, and the integration with primary healthcare systems.

AI-Powered Diagnostics

AI diagnostic tools that analyze medical images, laboratory results, and patient data are increasingly supporting clinical decision-making worldwide. These tools are particularly valuable in African settings, where there are few specialist physicians and limited diagnostic infrastructure. AI diagnostics are used in several critical areas:

- **Radiology:** AI systems interpret X-rays, CT scans, and ultrasounds to detect conditions such as tuberculosis (TB), pneumonia, and fractures. For example, Qure.ai's qXR system, deployed in hospitals across multiple African countries, achieved over 90% sensitivity for TB detection, often outperforming human radiologists in resource-limited settings (Qure.ai, 2023).

- **Pathology:** Artificial intelligence is being leveraged to analyze blood smears for the detection of malaria parasites, anemia, and other hematological conditions. These automated systems can reduce diagnostic time from hours to minutes and improve diagnostic consistency, which is particularly critical in high-volume settings.

- **Retinal Screening:** AI is widely used to detect diabetic retinopathy and other eye diseases in fundus photographs. This enables early intervention and blindness prevention, especially in countries with few ophthalmologists.

- **Infant Health:** Ubenwa, a Nigerian initiative, uses AI to analyze infant cries and detect birth asphyxia with 95% accuracy, facilitating timely interventions in resource-constrained birth settings.

- **Disease Prediction:** Machine learning models are increasingly used to forecast disease outbreaks, identify at-risk populations, and optimize resource allocation for diseases such as malaria and cholera, as well as emerging infectious diseases. The Africa Centres for Disease Control and Prevention (Africa CDC) is incorporating these predictive analytics into its early-warning systems.

Several challenges impede the advancement of AI-enabled diagnosis in African healthcare. These include the limited availability of Africa-specific data for training AI models; the potential for algorithmic bias when systems are trained primarily on non-African populations, which can reduce accuracy for African phenotypes; infrastructure limitations related to electricity and connectivity; difficulties in integrating AI into existing clinical workflows; and limited regulatory frameworks that ensure safety and efficacy without stifling innovation (Wahl et al., 2018). Addressing algorithmic bias requires developing locally trained AI models using African datasets, such as those created by initiatives like Lacuna Fund and Data Science Africa.

Box 6.2. Addressing Algorithmic Bias in Health AI

A significant challenge is that AI diagnostic tools trained on datasets from Europe and North America often underperform when applied to African populations. For example, skin lesion classifiers may exhibit significantly reduced accuracy on darker skin tones. To address this, initiatives such as the African AI research community, Masakhane's health datasets, and the Lacuna Fund's labelled data programme are working to build representative training corpora. Furthermore, the EU's AI Act (2024) provides a regulatory template that subject high-risk AI systems—including medical diagnostics—to demonstrate fairness across demographic groups before they can be deployed in the market—a legal framework that African countries can adapt.

Drones for Medical Delivery

Medical drones are overcoming geographic barriers by delivering essential supplies, including vaccines, blood products, medications, and emergency supplies, to remote and hard-to-reach areas. Zipline, which operates the world's largest autonomous drone delivery network, has been serving Rwanda since 2016 and Ghana since 2019, completing over 500,000 deliveries. Its drones operate autonomously, in all weather conditions, reaching health facilities within a 30-minute radius. Deliveries include blood products for postpartum hemorrhage (a leading cause of maternal mortality in sub-Saharan Africa), vaccines requiring cold storage, snake antivenoms, and COVID-19 diagnostic supplies (Scott & Scott, 2020).

In Ghana, a drone delivery network serves 12 million people across more than 2,000 healthcare facilities. Other African countries, including Nigeria, Kenya, Tanzania, Malawi, Ethiopia, and Côte d'Ivoire, are testing or planning similar drone-delivery systems for essential health supplies.

However, the long-term success of initiatives like Zipline faces challenges. These include establishing appropriate regulations and airspace management protocols, addressing weather-related issues in some areas, securing community support, and maintaining operational sustainability for programs that currently rely on donor or government support (Scott & Scott, 2020).

Box 6.3. Comparative Insight: Drone Delivery Beyond Africa

While Africa is at the forefront of medical drone delivery, other regions are also developing related models. India's Medicines from the Sky initiative in Telangana state uses drones to deliver vaccines and medications to remote tribal areas. In the Caribbean, Dominica has incorporated drone delivery for emergency medical supplies into its post-hurricane logistics strategy. The EU's Urban Air Mobility program is funding efforts to integrate drones into urban logistics networks. Africa's regulatory experience—particularly Ghana's progressive drone regulatory sandbox—is being studied globally as a potential model for balancing innovation and safety.

Health Data Analytics, Wearables, and Robotics

Health data analytics is becoming increasingly important in new technology applications within the health sector. The Africa CDC's response to COVID-19 highlighted the importance of integrated disease surveillance, real-time dashboards, and large-scale predictive modeling. In Kenya, Rwanda, South Africa, and Ethiopia, national health management information systems have adopted big-data analytics for epidemic monitoring, resource management, and supply chain oversight.

Wearable health devices, such as blood pressure monitors, glucose monitors, pulse oximeters, and maternal health trackers, are gaining popularity in Africa. However, their high cost and connectivity issues have mostly confined their use to urban, wealthier areas. In Nigeria, LifeBank employs IoT-connected blood banks and logistics platforms to enhance transparency in the blood supply chain and reduce shortages.

Medical robotics remains in early stages in Africa, mainly concentrated in tertiary hospitals across South Africa, Egypt, Kenya, and Nigeria. Current uses include robotic surgical assistance, rehabilitation robotics, and UV-disinfection robots, which were used during the COVID-19 pandemic. However, high costs, maintenance demands, and the need for specialized training limit their wider adoption.

Table 6.1: Synthesis of digital health solutions across Africa.

Technology	Key Applications	African Examples	Reach/ Impact	Key Challenges
Telemedicine	Remote consultations, specialist referrals, mental health	Babylon Health, mPharma, national platforms	22M consultation s/yr (2024)	Connectivity gaps, regulatory fragmentation, digital literacy
AI Diagnostics	Radiology, pathology, disease prediction, infant health	Qure.ai, Ubenwa, Jacaranda Health	Deployed in 15+ countries	Data scarcity, algorithmic bias, clinical integration
Drone Delivery	Vaccines, blood, medications, emergency supplies	Zipline (Rwanda, Ghana), Swoop Aero (Malawi)	500K+ deliveries, 12M+ served	Airspace regulation, cost sustainability, community acceptance
Health Analytics	Epidemic surveillance, resource optimisation	Africa CDC dashboards, national HMIS	Continental-scale deployment	Data quality, interoperability, technical capacity
Wearables/ IoT	Chronic disease management, maternal health	LifeBank, various pilot projects	Early adoption phase	Affordability, connectivity, data privacy
Robotics	Surgical assistance, rehabilitation, disinfection	Tertiary hospitals (SA, Egypt, Kenya)	Limited to major facilities	Costs, training, maintenance infrastructure

Sources: WHO (2022); Wahl et al. (2018); Scott & Scott (2020); Africa CDC (2024)

Digital health technologies hold significant potential to expand healthcare access, improve diagnostic accuracy, and reduce costs across Africa. However, realizing their potential depends on coordinated investments in connectivity infrastructure, harmonized regulatory frameworks, training for health workers on digital tools, robust data governance practices, and sustainable funding models. Integrating these solutions with existing healthcare systems is crucial; digital solutions should complement, not replace, existing structures. Furthermore, to ensure equitable access, strategies must deliberately target marginalized populations, including those in rural areas, informal settlements, and conflict zones (Wahl et al., 2018).

6.3 Education, Skills, and Learning Technologies

Africa's education systems face complex, interconnected challenges. About 98 million school-age children and adolescents are out of school. Learning outcomes are generally poor. Chronic teacher shortages persist, and there is a significant skills mismatch between educational outcomes and the requirements of rapidly changing labor markets (UNESCO, 2023).

Educational technologies (EdTech), including digital learning platforms, adaptive learning systems, immersive technologies, and digital skills training, offer new ways to improve learning outcomes, increase access to education, and prepare young people for careers in the digital economy.

Digital Learning Platforms

Digital learning platforms leverage emerging technologies to deliver educational content on mobile phones, tablets, or computers. These platforms support self-paced learning, supplementary instruction, and alternative education options. There are African EdTech companies that have emerged to address specific local needs. These include:

- **Eneza Education** (Kenya, Ghana, Côte d'Ivoire): This platform uses SMS and mobile web to deliver curriculum-aligned content for primary and secondary students. Over 10 million learners access its free and paid services. Its simple design (SMS and 2G compatibility) ensures accessibility in areas with limited connectivity.

- **Ubongo (Tanzania, pan-African):** This educational entertainment program reaches 34 million children across more than 40 countries through television, radio, and digital platforms. It offers localized content in African languages, including Swahili, Kinyarwanda, Hausa, and others. The app teaches literacy, numeracy, and life skills.

- **M-Shule (Kenya)**: This AI-powered learning platform uses SMS and mobile web to provide personalized lessons. Adaptive algorithms adjust the difficulty of content based on student performance data.

Evidence concerning the learning outcomes of EdTech programs is mixed. Well-implemented programs with strong teacher support, curriculum alignment, and adaptive features have demonstrated learning gains of 15–25%. However, many deployments yield minimal impact due to suboptimal implementation, inadequate teacher training, lack of curriculum integration, or technology failures.

The data in Figure 6.2 reveal a trade-off between scale and impact in technology-enabled learning. While widely adopted solutions, such as mobile learning, reach a broad audience, they often lead to moderate learning gains.

Figure 6.2: Reach and estimated learning impact of EdTech Solutions

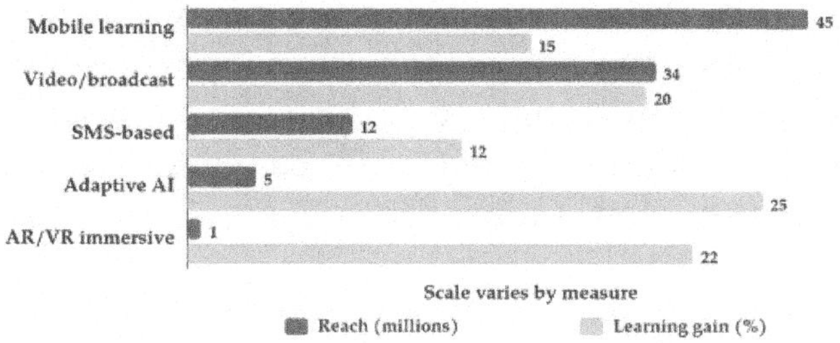

Sources: UNESCO (2023); Pitan & Muller (2019)

Personalized or immersive technologies offer greater potential impact but face scalability challenges across the continent. Additionally, evidence shows that technology alone is not enough to ensure improved learning outcomes; effective pedagogical design and systemic integration are essential (UNESCO, 2023).

Immersive Technologies: Augmented and Virtual Reality

Augmented reality (AR) and virtual reality (VR) offer immersive learning experiences, such as virtual science labs, historical site explorations, surgical simulations, and engineering design environments. These technologies are especially useful when physical facilities are limited, unsafe, or too expensive. For example, virtual chemistry labs eliminate the need for expensive equipment and hazardous materials. Medical training simulations allow surgical practice without risking patient safety, and virtual field trips can take students to African heritage sites and natural environments (Jensen & Konradsen, 2018).

However, several factors limit the adoption of AR/VR in African education, including high hardware costs (USD 200–1,000 per VR headset), bandwidth requirements for streaming immersive content, and the cost of developing locally relevant educational content. As a result, most AR/VR applications remain in the pilot stage rather than achieving widespread use. More affordable options, such as Google Cardboard, smartphone-compatible VR, and tablet-based AR, offer lower-cost entry points but lower immersion quality.

Box 6.4 Comparative Insight: Immersive Education in the EU and Asia

The European Commission's Digital Education Action Plan (2021–2027) funds immersive learning pilots across its Member States and invests significantly in virtual laboratories for STEM education. South Korea's national curriculum incorporates VR content into history and science classes in public schools, supported by the distribution of government-subsidized VR hardware. China has deployed VR classrooms in over 10,000 schools, particularly in rural areas with a shortage of qualified teachers. These examples suggest that government-led investment and curriculum integration are necessary to expand immersive education beyond pilot projects.

Skills Training for the Digital Economy

Digital skills training prepares young people for careers in the technology sector and the broader digital economy. A vibrant network of training programs is evolving in Africa through:

- **Coding bootcamps** are intensive programs that teach programming, web development, data science, and software engineering. Andela has trained over 100,000 developers across Africa. Moringa School (Kenya) offers immersive full-stack development programs, and ALX Africa provides technology leadership and software engineering training at scale.

- **Online apprenticeships and remote work platforms:** Andela pioneered a model that connects African developers with global technology companies. Similar models are now emerging for design, digital marketing, data annotation, and other digital services.

- **Micro-credentials and professional certificates:** Short courses that offer industry-recognized credentials, such as Google Career Certificates, Coursera Professional Certificates, and AWS certifications, provide affordable ways to gain digital skills.

- **Government ICT programs:** National initiatives like Rwanda's ICT in Education program, Kenya's Digital Literacy Programme, Nigeria's Digital Nigeria initiative, and Ethiopia's five million coders program aim to promote widespread digital literacy and skills for employment in the tech industry.

However, employment outcomes from these initiatives vary widely across countries. While top graduates from intensive training programs may secure competitive positions in the tech industry, most employment programs only reach a small fraction of the youth population. The quality of training is often inconsistent. Scalability is also limited by instructor availability, infrastructure needs, student financing, and alignment with employers' specific requirements (Pitan & Muller, 2019).

The Digital Divide in Education

A major obstacle to wider adoption of technology-enabled education is the digital divide. Data indicate a substantial 73-percentage-point gap between the most privileged groups (urban, higher-income) and the most disadvantaged ones (rural, lower-income), illustrating how income and location together affect access. The clear downward trend indicates that digital access systematically decreases across both urban-rural and wealth spectrums. The COVID-19 pandemic highlighted these disparities, as students with digital access continued learning online, while those without experienced significant educational setbacks (UNESCO, 2023).

Figure 6.3: Access to digital learning resources by location and income group in Africa

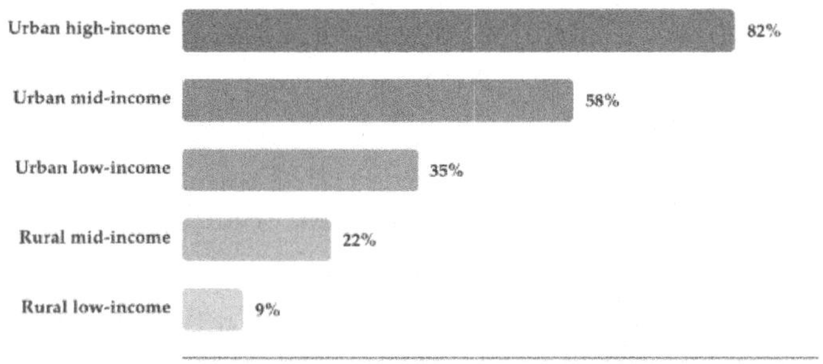

.Source: UNESCO (2023)

Without targeted equity strategies—such as subsidized devices and connectivity, offline-capable solutions, community access points, and multilingual content—educational technology risks worsening existing inequalities. Addressing education's emerging technology adoption gaps requires recognizing connectivity and devices as essential parts of the educational infrastructure, on par with classrooms and textbooks, and developing solutions that work well in low-bandwidth, low-resource environments.

6.4 Agriculture, Food Security, and Climate Adaptation

Agriculture is a crucial sector in Africa, employing about 60% of the workforce and contributing between 15% and 35% to most countries' GDP. However, productivity remains generally low, with cereal yields roughly one-third of the global average. The sector is more vulnerable to climate change, with rising temperatures, unpredictable rainfall, and extreme weather events endangering food security for hundreds of millions (Pretty et al., 2018). Agricultural technologies, including precision agriculture, climate-smart practices, market access platforms, digital finance, and IoT solutions, have the potential to boost productivity, strengthen climate resilience, improve market access, and increase farmers' incomes.

Precision Agriculture

Precision agriculture uses data and technology to optimize the use of inputs—seeds, fertilizers, water, and pesticides—based on specific field conditions, offering a precise alternative to uniform application. This approach has the potential to boost yields, lower input costs, and minimize environmental impact. Emerging technology-driven implementations in African contexts include:

- **Soil sensors and IoT**: Internet-connected sensors measure soil moisture, nutrient levels, pH, and temperature, enabling targeted irrigation and fertilization strategies. Studies show that precision methods can reduce input costs by 20–30% while maintaining or increasing yields (Tey & Brindal, 2012).

- **Satellite imagery and remote sensing**: Earth observation platforms monitor crop health, identify stressed areas, forecast yields, and detect pest and disease outbreaks.

Complimentary imagery from the European Space Agency's Sentinel constellation and NASA's Landsat program increases access, while commercial high-resolution providers offer advanced capabilities for larger operations.

- **Hyperlocal weather forecasting:** Platforms like Ignitia in West Africa provide SMS-based forecasts with 84% accuracy, significantly outperforming traditional models in tropical regions (around 39% accuracy).

- **Variable-rate application:** Precision equipment adjusts seed, fertilizer, or pesticide application rates based on within-field variability. Emerging service models, such as Hello Tractor's equipment-sharing platform, extend mechanization opportunities to smallholders.

Despite its potential to significantly improve agricultural productivity, the adoption of precision agriculture in Africa remains limited. Currently, only a small fraction of smallholder farmers have access to these tools.

Barriers include costs, technical complexity, connectivity requirements in rural areas, and limited demonstrated return on investment for smallholder farming systems. However, simpler mobile-based advisory services that combine weather data, agronomic advice, and market prices already reach an estimated 12.5 million farmers, demonstrating the scalability of lower-tech approaches (Tey & Brindal, 2012).

Market Access Platforms

Digital platforms connect farmers to markets, improving price transparency, reducing post-harvest losses—which account for 30–40% of production in sub-Saharan Africa—and increasing incomes. For example, Twiga Foods in Kenya, a business-to-business (B2B) platform, links farmers to urban retailers through mobile ordering, logistics networks, and mobile money payments. This platform supports over 17,000 farmers and 50,000 retail outlets, with farmers earning 15–20% higher prices compared to traditional intermediary channels.

WeFarm, another example operating in Kenya and Uganda, is a peer-to-peer knowledge-sharing platform that allows farmers to ask questions and share solutions via SMS, serving 3.5 million users. These platforms provide price information services by transmitting commodity prices across markets via SMS. This helps reduce information asymmetries that intermediaries usually exploit, enabling participating farmers to receive 10–15% higher prices for their crops.

Box 6.5. Comparative Insight: AgriTech in Asia and Latin America

India's e-National Agriculture Market (eNAM) digitally connects over 1,000 wholesale produce markets, facilitating price discovery and reducing intermediary margins. In China, the Pinduoduo platform has connected over 16 million farmers with urban consumers through group-buying models. In Latin America, Colombia's AgroNet platform provides market intelligence to smallholder coffee and cocoa farmers, while Brazil's Embrapa agricultural research system provides digital advisory services to farmer cooperatives.

Agricultural Financing and Insurance

Traditional financial institutions often avoid smallholder farmers because of perceived risks and high transaction costs. However, digital finance innovations are beginning to overcome these barriers. Technology-driven digital lending platforms, such as Apollo Agriculture in Kenya and FarmCrowdy in

Nigeria, offer credit for inputs that are delivered directly to farms. These platforms use alternative data—such as mobile phone usage patterns, sales history, and satellite imagery—to evaluate creditworthiness. Repayments are made at harvest via mobile money. Collectively, these platforms serve an estimated 5.5 million farmers.

Index-based crop insurance relies on satellite data or weather station measurements to automatically trigger payouts when certain thresholds are met, reducing the need for costly field assessments. However, data indicate that basis risk—the gap between insurance triggers and actual on-farm losses—limits farmer confidence and adoption (Pretty et al., 2018).

Climate Change Adaptation

Emerging technologies support climate change adaptation in several ways. Genomic breeding accelerates the development of crop varieties that are drought-resistant, heat-tolerant, and pest-resistant. Climate forecasting systems enable proactive actions, allowing farmers to adjust planting dates, choose suitable varieties, and prepare for extreme weather events. Precision irrigation systems optimize water use efficiency in water-scarce regions. Data analytics support diversification strategies, helping farmers in adjusting their cropping patterns to match changing conditions.

Despite these advances, many smallholder farmers still cannot access most climate adaptation technologies due to costs, complexity, and distribution challenges. Scaling up climate-smart practices requires combining indigenous and traditional knowledge with modern technologies, supporting farmer-led experimentation and peer learning, and mobilizing dedicated climate finance for agricultural adaptation (Pretty et al., 2018).

Table 6.2: AgriTech solution categories, reach, impact, and challenges

.

Solution Category	African Examples	Farmers Reached	Measured Impact	Challenges
Weather & Advisory	Ignitia, FarmRadio, extension apps	12.5M farmers	15–20% yield gains	User retention, sustainable funding models
Market Platforms	Twiga Foods, WeFarm, price SMS	8.2M farmers	15–20% higher prices	Logistics costs, trust, and post-harvest losses
Digital Credit	Apollo Agriculture, FarmCrowdy	5.5M farmers	30–40% yield increase	Default management, data quality, scale
Input & Equipment	Hello Tractor, equipment sharing	6.8M farmers	20% cost reduction	Asset maintenance, coordination, and training
IoT Precision	Soil sensors, smart irrigation	1.2M farmers	25–35% input savings	Costs, connectivity, technical complexity
Climate Adaptation	Genomic breeding, index insurance	3M+ farmers	Reduced climate losses	Basis risk, seed access, distribution

Sources: Pretty et al. (2018); Tey & Brindal (2012).

Although digital agriculture solutions are expanding and delivering measurable gains in productivity and income, challenges related to financial sustainability, infrastructure, coordination, and risk management still hinder their scalability and long-term success.

6.5 Urban Development, Smart Cities, and Transport

Africa is undergoing rapid urbanization, with cities growing at about 3.5% annually. The urban population is projected to triple by 2050, reaching around 1.5 billion people. This rapid growth brings significant challenges, including traffic congestion, pollution, insufficient infrastructure, overloaded services, and the spread of informal settlements, as well as opportunities for economic growth, innovation, and improved service delivery (Odendaal, 2016).

Figure 6.4. Africa's Urban and Rural Population Trajectories (2000–2050)

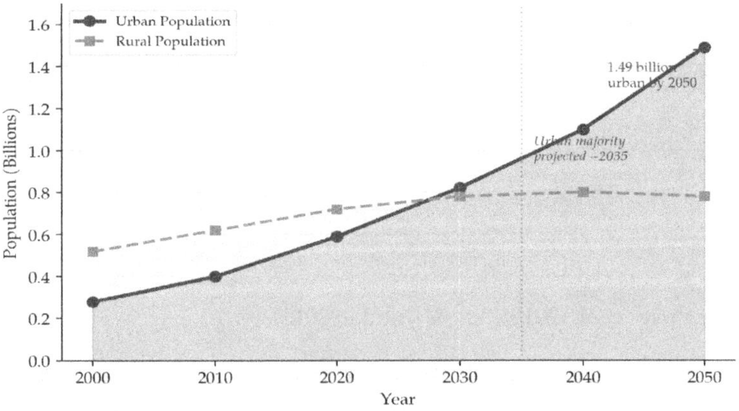

Source: UN DESA (2022).

Intelligent Transportation Systems

Transportation is the main challenge of urban life in Africa. In many major cities, average commute times exceed 90 minutes; traffic congestion causes estimated GDP losses of 2–10%; and road traffic accident rates are among the highest globally. Emerging technology solutions are being embraced across different aspects of urban life, including:

- **Ride-sharing and mobility platforms:** Companies like Uber and Bolt provide on-demand transportation, along with local services including SafeMotos (Rwanda), SafeBoda (Uganda), and Gokada (Nigeria). These apps have digitalized motorcycle taxis, enhancing safety through rider screening and GPS tracking.

- **Public transit optimization:** Mobile apps now provide real-time transit updates, digital ticketing, and route planning. Digital systems are being integrated into Lagos' Bus Rapid Transit (BRT), Nairobi's matatu routes, and Dar es Salaam's rapid transit.

- **Traffic Management:** Cities such as Kigali, Nairobi, and Johannesburg are using sensors, cameras, and data analytics to improve traffic flow. However, most African cities still lack comprehensive intelligent transportation systems.

- **Last-Mile Delivery and Logistics:** Platforms like Sendy, Kobo360, and Lori Systems are improving delivery routes, tracking packages, and coordinating couriers, all of which are essential for e-commerce growth.

Box 6.6 Comparative Insight: Smart Mobility in Asia

Singapore's intelligent transport system manages congestion through electronic road pricing, comprehensive sensor networks, and real-time data analytics. Southeast Asian super-apps like Grab and GoTo integrate ride-sharing, delivery, and payments on a single platform, a model that African platforms are beginning to emulate.

Smart Utilities and Resource Management

Smart utility systems are essential for optimizing resource use and enhancing service reliability in African cities. Smart water systems that use sensors to detect leaks—reducing water loss by 20–40%—monitor water quality in real time and support demand management. Nairobi, Cape Town, and Dakar are implementing smart water systems at different levels.

To improve grid efficiency, smart electricity grids incorporate advanced metering infrastructure, demand response mechanisms, and integrated distributed generation. In waste management, digital platforms like Pakam in Cameroon and WeCyclers in Nigeria optimize collection routes with fill-level sensors and GPS tracking, while promoting recycling through incentive-based programs.

Digital Governance and Citizen Services

Digital governance platforms offer significant opportunities to improve municipal service delivery and citizen engagement. Online portals can streamline business registration, permit applications, tax payments, and service requests. Mobile apps enable citizens to report infrastructure issues, request services, and provide feedback to city officials.

Digital city projects in Kigali, open data portals in Cape Town, and digital service platforms in Nairobi illustrate how technology can improve efficiency and transparency in urban governance. However, many African cities still face limited e-governance capabilities. Challenges such as digital literacy barriers (which risk excluding marginalized groups), poor infrastructure, bureaucratic resistance to change, and cybersecurity threats constrain their wider adoption (Odendaal, 2016).

Table 6.3: Smart city technology domains, applications, and outcomes across African cities.

Domain	Technologies	African Examples	Outcomes	Barriers
Transportation	Ride-sharing, transit apps, traffic mgmt	Uber, SafeMotos, BRT systems	20–30% time savings	Regulation, infrastructure
Water	Smart meters, leak detection, quality	Nairobi, Cape Town, Dakar	20–40% loss reduction	Investment, maintenance
Energy	Smart grids, AMI, demand response	South Africa, Kenya	10–15% efficiency gains	Grid modernization costs
Waste	Route optimization, recycling tracking	Pakam, WeCyclers	Cost savings, recycling up	Informal sector integration
Governance	E-services, mobile apps, open data	Kigali, Cape Town, Nairobi	Faster services	Digital divide, bureaucracy
Environment	Air quality, pollution tracking	City-level sensor networks	Policy-relevant data	Sensor maintenance

Sources: Odendaal (2016); ITU (2023).

Applying emerging technologies in African cities faces multiple challenges that must be addressed early on. Most smart city initiatives focus more on technology than on people, emphasizing technical solutions over participatory governance. These projects often benefit wealthier urban residents, leaving informal settlements and marginalized groups behind. The surveillance systems embedded in some smart city setups can threaten privacy and civil liberties. Additionally, the high, often unsustainable costs of these projects can create long-term dependence on vendors.

For successful smart city development in the African context, it is essential to include all residents in the planning and governance processes, use technology tailored to local needs and capacities, establish sustainable financing models, implement robust data governance frameworks, and integrate with existing urban systems (Odendaal, 2016).

6.6 Cross-Cutting Challenges and Enabling Conditions

The cross-cutting challenges and enabling conditions for the adoption of emerging technologies in health, education, agriculture, and urban development include infrastructure gaps, skills shortages, and digital inequality.

Infrastructure deficits, such as unreliable electricity, limited broadband coverage, and high data costs, further constrain digital adoption and innovation across these sectors and others. Addressing these challenges requires policies that expand universal connectivity, invest in digital public infrastructure, promote infrastructure sharing, and deploy renewable energy and satellite solutions to extend connectivity to underserved areas.

Many sectoral digital initiatives face sustainability challenges due to donor dependence, fluctuating investment flows, and weak revenue models. At the same time, regulatory gaps, limited algorithmic accountability, shortages of digital skills, and fragmented digital systems restrict scaling and impact.

Table 6.4: Cross-cutting challenges and enabling conditions for technology-driven development.

Challenge Area	Description	Enabling Conditions
Digital Divides & Equity	Urban–rural, gender, income, age, and disability-based disparities in technology access, use, and benefit	Universal access policies, subsidised connectivity, inclusive design, multilingual content, gender-responsive strategies
Infrastructure Deficits	Unreliable electricity, limited broadband, high data costs, and inadequate physical infrastructure	Investment in digital public infrastructure, renewable energy, satellite connectivity, infrastructure sharing
Sustainability & Financing	Donor dependency, venture capital volatility, unclear revenue models, cost recovery challenges	Blended finance, public–private partnerships, results-based financing, social enterprise models
Governance & Regulation	Data protection gaps, algorithmic accountability deficits, regulatory fragmentation	Continental frameworks, regulatory sandboxes, adaptive regulation, capacity building for regulators
Human Capital	Digital literacy gaps, STEM skill shortages, brain drain, gender gaps in technology fields	Expanded STEM education, digital literacy programmes, diaspora engagement, reskilling and upskilling
Integration	Parallel systems, weak interoperability, poor linkages with existing institutions	Open standards, digital public goods, government-led coordination, iterative scaling

Source: Authors' synthesis.

Enabling conditions that address these challenges include blended finance and public–private partnerships, stronger continental regulatory frameworks, regulatory sandboxes, expanded STEM and digital literacy programs, and the adoption of open standards and interoperable digital public goods that ensure coordinated and sustainable technology deployment.

6.7 Conclusion

Emerging technologies hold significant potential for growth in Africa and the Global South, especially in health, education, agriculture, and urban systems. Digital health innovations are expanding access to healthcare for underserved communities, while educational technologies are improving learning outcomes and opening more opportunities for young people. In agriculture, innovations are improving productivity and climate resilience. Smart city solutions are addressing the challenges of rapid urbanization. Collectively, these advancements can accelerate progress toward the Sustainable Development Goals and Agenda 2063.

However, technology alone is not sufficient to guarantee developmental progress. Unlocking the transformative potential of new technologies requires addressing ongoing structural issues. These issues include digital divides that may deepen inequalities, infrastructure gaps in connectivity and electricity that hinder deployment, uncertainties related to sustainability in business models and funding, challenges in integrating technologies as part of broader systems rather than as separate entities, governance frameworks that fall behind innovation, and evidence gaps that hinder accurate impact assessments.

Experiences from Asia, Europe, Latin America and Oceania reinforce key lessons: government-led investment in digital public infrastructure and regulatory frameworks is essential; integration with existing systems and institutions is more important than technological novelty; and inclusive design that prioritizes marginalized populations is necessary for fair outcomes. Success requires moving beyond simple enthusiasm for technology to evidence-based deployment that emphasizes equity, sustainability, and systemic integration.

The next chapter further explores economic and productive applications, showing how technologies are transforming industries, supporting small and medium enterprises, enabling digital trade, and reshaping the future of work. Together, social and economic applications demonstrate the full scope of technology's transformative potential in Africa, while also highlighting the governance, financing, and implementation challenges that require strategic focus from policymakers, practitioners, and development partners across the continent.

References

Africa CDC. (2024). Digital health transformation: Annual report 2023–2024. Africa Centres for Disease Control and Prevention.

African Union. (2020). The digital transformation strategy for Africa (2020–2030). African Union Commission.

Government of India. (2023). eSanjeevani: National telemedicine service annual report. Ministry of Health and Family Welfare.

Grand View Research. (2024). Digital health market size, share and trends analysis report, 2024–2030. Grand View Research.

Harzheim, E., Gonçalves, M. R., Umpierre, R. N., da Silva Siqueira, A. C., Katz, N., Agostinho, M. R., et al. (2019). Telehealth in Rio Grande do Sul, Brazil: Bridging the gaps. Telemedicine and e-Health, 22(11), 938–944.

Jensen, L., & Konradsen, F. (2018). A review of the use of virtual reality head-mounted displays in education and training. Education and Information Technologies, 23(4), 1515–1529.

Mars, M. (2020). Telemedicine and advances in urban and rural healthcare delivery in Africa. Progress in Cardiovascular Diseases, 63(3), 326–335.

Munung, N. S., de Vries, J., et al. (2022). Genomics research ethics in Africa: Evolving frameworks. Nature Genetics, 54(1), 12–18.

Odendaal, N. (2016). Smart city: Neoliberal discourse or urban development tool? In S. Marvin, A. Luque-Ayala, & C. McFarlane (Eds.), Smart urbanism: Utopian vision or false dawn? (pp. 33–51). Routledge.

Pitan, O. S., & Muller, C. (2019). University reputation and undergraduate employability in the labour market. Industry and Higher Education, 33(5), 331–341.

Pretty, J., Benton, T. G., Bharucha, Z. P., Dicks, L. V., Flora, C. B., Godfray, H. C. J., et al. (2018). Global assessment of agricultural system redesign for sustainable intensification. Nature Sustainability, 1(8), 441–446.

Qure.ai. (2023). qXR: AI-powered chest X-ray solution—Deployment report. Qure.ai Technologies.

Sachs, J., Schmidt-Traub, G., Kroll, C., Lafortune, G., Fuller, G., & Woelm, F. (2019). Sustainable development report 2019. Bertelsmann Stiftung und SDSN.

Scott, J. E., & Scott, C. H. (2020). Drone delivery models for healthcare, in Proceedings of the 53rd Hawaii International Conference on System Sciences (pp. 3297–3306).

Tey, Y. S., & Brindal, M. (2012). Factors influencing the adoption of precision agricultural technologies. Precision Agriculture, 13(6), 713–730.

UNESCO. (2023). Global education monitoring report 2023: Technology in education—A tool on whose terms? UNESCO Publishing.

Wahl, B., Cossy-Gantner, A., Germann, S., & Schwalbe, N. R. (2018). Artificial intelligence (AI) and global health. BMJ Global Health, 3(4), e000798.

World Health Organization. (2022). World health statistics 2022: Monitoring health for the SDGs. WHO.

Chapter 7: Economic and Productive Applications of Emerging Technologies

7.1 Introduction

Chapter 6 explored the social aspects of deploying emerging technologies, focusing on health, education, agriculture, and urban systems. This chapter shifts to the economic and productive sphere, exploring how these technologies are reshaping industrial production across Africa.

Emerging technologies are changing the operating environment for small and medium enterprises and startups, enabling new types of digital trade and platform-based services, and fundamentally transforming the nature and distribution of work. These changes have profound implications for the continent's aspirations for structural transformation, its integration into global value chains, and the livelihoods of hundreds of millions of people (African Development Bank, 2024; World Economic Forum, 2023).

This chapter is organized into four sections. Section 7.2 examines industrialization and smart manufacturing, including the potential of Industry 4.0 technologies to accelerate Africa's manufacturing renaissance. Section 7.3 explores the vibrant SME and start-up ecosystem, assessing how digital tools expand market access, improve productivity, and attract new forms of financing. Section 7.4 analyzes the growth of digital trade, platform economies, and creative industries. Section 7.5 confronts the dual impact of technology on employment—job creation alongside displacement—and the evolving skills landscape. Throughout, the analysis draws on comparative experiences from the Global South to contextualize African developments and identify transferable lessons.

A central argument in this chapter is that while technology offers powerful tools for economic transformation, how its benefits are shared depends on complementary investments in human capital, infrastructure, institutional frameworks, and intentional policies that promote inclusion. Without these investments, technological change risks exacerbating existing inequalities—between countries, firms, and workers—rather than reducing them (UNCTAD, 2023).

7.2 Industrialization and Smart Manufacturing

Africa's manufacturing sector accounts for approximately 11% of continental GDP, significantly below the 20–35% observed in East and Southeast Asia. Furthermore, its share of output has stagnated or declined in several major African economies since the 1990s (African Union & UNIDO, 2022). This premature deindustrialization, coupled with heavy dependence on commodity exports and limited participation in global value chains, highlights the urgent need for industrial renewal. Industry 4.0 technologies—including smart manufacturing, additive manufacturing, robotics, digital twins, and supply chain digitization—present both opportunities and challenges in pursuing this ambition.

There are six areas where emerging technologies can be applied in manufacturing:

- **Smart Manufacturing:** This involves automation and the use of the Internet of Things to drive data-enabled production processes and real-time monitoring.

- **Green Manufacturing:** This focuses on efficiency and circular production models to support sustainability.

- **Supply Chain Digitization:** This aims to enhance tracking, quality control, and transparency across value chains.

- **Robotics and Automation:** These are applied to strengthen assembly and packaging capabilities, improving productivity and consistency.

- **Digital Twins:** These enable the simulation and optimization of industrial processes before physical deployment.

- **3D Printing and Local Production:** These support distributed manufacturing of medical supplies and spare parts, reducing dependence on imports.

Smart Manufacturing and Industry 4.0

Smart manufacturing integrates IoT sensors, automation, artificial intelligence, and data analytics in production processes. Relevant applications include: predictive maintenance, which uses sensor-based condition monitoring to reduce downtime by 30–50% and maintenance costs by 10–25%; quality control, where computer vision systems detect defects in real time, reducing waste and improving consistency; and production optimization, which uses AI algorithms to plan production schedules, manage energy consumption, and minimize material waste. Digital twin technology, which creates virtual models of physical systems, allows manufacturers to simulate processes, test changes, and improve operations before implementing them on the factory floor.

Smart manufacturing adoption is limited to a few countries and sectors in Africa. Early adopters include South Africa's automotive industry, Egypt's textile and garment sector, Morocco's aerospace components manufacturing, and Kenya's agro-processing industries.

In South Africa, the manufacturing sector has piloted sensor-equipped production lines in the automotive and mining sectors, demonstrating cost savings of 15–20% through predictive maintenance alone (UNIDO, 2022). Egyptian garment factories have implemented computer vision for quality inspection, improving defect detection rates and reducing return rates for European buyers.

Figure 7.1 shows the current Manufacturing Value Added (MVA) as a percentage of GDP, along with the projected MVA under Industry 4.0 adoption, for select African economies: South Africa, Egypt, Nigeria, Morocco, Kenya, Ethiopia, Tanzania, and Ghana.

The chart shows a consistent upward trend in manufacturing's contribution to GDP, driven by the Industry 4.0 transformation. For example, South Africa's MVA increases from about 12% to 18%, Egypt's from roughly 17% to 22%, Nigeria's from 9% to 15%, and Morocco's from 15% to 20%. Emerging manufacturing economies such as Kenya, Ethiopia, Tanzania, and Ghana also show significant potential gains, with projected increases of 4–7 percentage points.

Figure 7.1 Manufacturing value added as a percentage of GDP—current levels and projected potential with Industry 4.0 adoption.

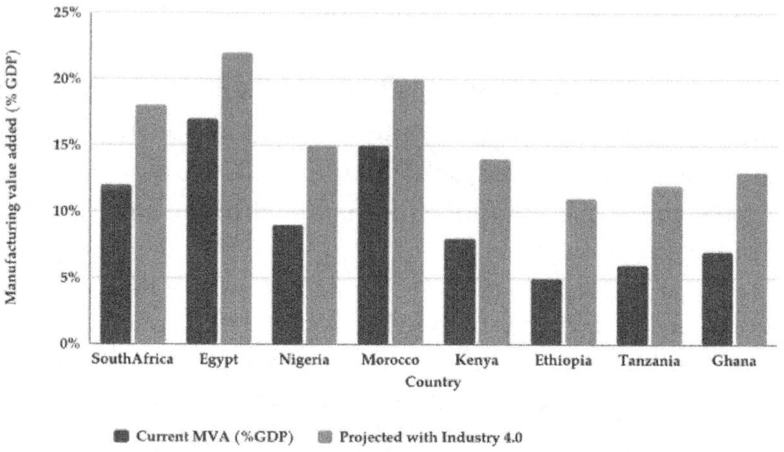

Sources: UNIDO (2022); AfDB (2024).

However, several barriers hinder the wider adoption of smart manufacturing. Reliable electricity and broadband connectivity are still lacking in many industrial zones. The upfront capital costs of sensor networks, automation equipment, and software platforms are too high for most companies, especially SMEs.

Additionally, there is a shortage of technical skills in industrial IoT, data science, and automation engineering. Furthermore, the business case for Industry 4.0 investments is often unclear for firms operating in low-wage settings, where the economic rationale for automation differs significantly from those in high-wage economies (UNIDO, 2022).

Box 7.1 Comparative Insight: Industry 4.0 in Asia and Latin America

China's 'Made in China 2025' strategy invested over USD 300 billion in smart manufacturing, propelling the country to global leadership in deploying industrial robotics. Vietnam and Bangladesh have integrated IoT-based quality monitoring in garment factories to meet European buyer standards. In Latin America, Mexico's automotive sector has adopted digital twin technology in partnership with German OEMs, while Brazil's SENAI network provides Industry 4.0 training to over 500,000 manufacturing workers annually. The key lesson is that successful industrial digitization requires coordinated ecosystem support combining infrastructure investment, skills development, and technology financing, rather than relying solely on firm-level adoption.

Additive Manufacturing (3D Printing)

Additive manufacturing enables localized production, eliminating the need for expensive molds, tooling, and large inventories. A number of uses are especially relevant to Africa, including the production of medical devices such as prosthetics, orthotics, surgical instruments, and anatomical models for surgical planning.

3D printing enables local manufacturing of replacement components for machinery and vehicles while reducing reliance on costly, slow international supply chains. Building, as demonstrated by 3D-printed housing prototypes in countries such as Kenya and Malawi, offers a faster, cheaper construction option with less material waste.

In Kenya, university and innovation hub 3D printing labs are producing affordable prosthetic limbs at a fraction of the cost of imported models. In South Africa, the Council for Scientific and

Industrial Research has developed 3D-printed titanium implants for reconstructive surgeries. Nigerian start-ups are using 3D printing to produce spare parts for agricultural machinery, reducing farmer downtime from weeks (waiting for imported parts) to days. These applications address a critical issue: Africa imports over USD 50 billion in manufactured goods annually, much of which could be produced locally using additive manufacturing technologies (AfDB, 2024).

However, scaling additive manufacturing faces obstacles. Challenges include the high cost of industrial-grade 3D printers (USD 50,000–1,000,000), limited access to printing materials (many of which must be imported), difficulties in quality assurance and certification, and a shortage of skilled operators and designers. Despite these challenges, as printer costs decrease and material options expand, additive manufacturing is set to play an increasingly important role in African industrial strategies.

Green Manufacturing and Circular Economy

Emerging technologies support greener manufacturing practices. For example, energy management systems powered by IoT sensors and AI can reduce energy consumption in factories. Waste reduction through precision manufacturing and recycling supports the principles of the circular economy. Additionally, carbon-tracking and reporting tools, which are increasingly required by international buyers and regulators, enable African manufacturers to demonstrate environmental compliance and access green financing.

South Africa's industrial energy efficiency program, implemented in partnership with the United Nations Industrial Development Organization, has saved over 9,000 GWh of energy across participating industries. Ethiopia's industrial parks—such as Hawassa, Bole Lemi, and Kilinto—are incorporating renewable energy, water recycling, and waste treatment into their designs, serving as models for green industrial zones.

Table 7.1 Industry 4.0 technologies, applications, adoption stages, and barriers in African manufacturing.

Technology	Key Applications	African Examples	Adoption Stage	Key Barriers
Smart Manufacturing	Predictive maintenance, quality control, production optimisation	SA automotive, Egypt textiles, Kenya agro-processing	Early adoption in select sectors	Electricity, connectivity, capital costs, skills
3D Printing	Medical devices, spare parts, construction prototypes	Kenya prosthetics, SA implants, Nigeria parts	Innovation hubs, university labs	Printer/ material costs, certification, skills
Digital Twins	Process simulation, optimisation, training	Mining (SA), manufacturing pilots	Pilot stage	Data infrastructure, software licences, expertise
Robotics	Assembly, packaging, welding, and painting	SA automotive, Egypt manufacturing	Large firms only	Capital costs, job displacement concerns, and maintenance
Supply Chain Tech	Tracking, traceability, quality assurance	Export agriculture, minerals (pan-African)	Growing adoption	Interoperability, standards, cost
Green Manufacturing	Energy efficiency, waste reduction, and carbon tracking	SA energy programme, Ethiopia industrial parks	Policy-driven adoption	Investment, measurement, market incentives

Sources: UNIDO (2022); AfDB (2024).

Industry 4.0 technologies are present across Africa, but their widespread adoption hinges on overcoming structural barriers related to infrastructure, financing, regulation, and human capital.

7.3 SMEs, Start-ups, and Entrepreneurship

Small and medium enterprises (SMEs) are the backbone of African economies, representing approximately 80–90% of all businesses, 50–60% of employment, and 40–50% of GDP in most countries (International Finance Corporation, 2023). However, most African SMEs operate informally, with limited access to finance, markets, technology, and business services.

Emerging technologies are transforming this landscape by expanding market access, improving operational efficiency, enabling new financing models, and connecting entrepreneurs to global knowledge networks. At the same time, a vibrant start-up ecosystem has emerged, attracting substantial venture capital and creating innovative solutions to African challenges.

Figure 7.2 shows venture capital funding trends for African start-ups from 2017 to 2024, measured in billions of U.S. dollars. Investment steadily grew from $0.6 billion in 2017 to $2.0 billion in 2019, then slightly declined to $1.4 billion in 2020, due to pandemic-related market uncertainty. A surge in funding followed during the post-pandemic expansion, reaching $5.2 billion in 2021 and peaking at $6.5 billion in 2022. The data shows a significant market correction in 2023, with funding falling to $3.2 billion, reflecting tighter global financial conditions and reduced venture activity. However, 2024 shows signs of recovery, with funding rebounding to $3.8 billion, suggesting renewed investor confidence but still below the highs of 2021–2022.

Figure 7.2 Venture capital funding for African start-ups, 2017–2024, showing the growth trend, 2021–2022 peak, and subsequent market correction.

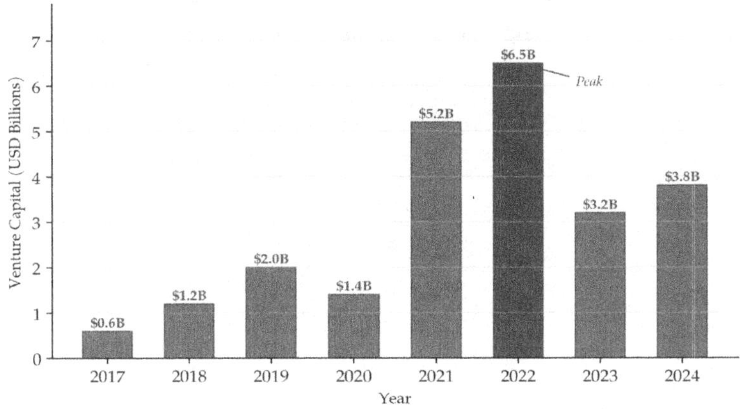

Sources: Partech Africa (2024); Briter Bridges (2024).

The chart highlights both the rapid growth and the volatility of Africa's venture capital scene, showing increasing global interest in African innovation ecosystems while also revealing sensitivity to global macroeconomic cycles.

Digital Business Tools for SMEs

Digital tools enhance SME productivity and competitiveness. Mobile money and digital payments—adopted by about 68% of African SMEs in countries with mature mobile money systems—reduce transaction costs, improve cash flow management, and build digital financial footprints that help secure credit. Social media marketing provides an affordable way to reach customers. Digital accounting and inventory management tools promote financial discipline and transparency. E-commerce platforms expand market reach beyond local areas (IFC, 2023).

However, a significant digital divide persists among SMEs. Firms in capitals and major cities, led by younger and more educated entrepreneurs, adopt digital tools at much higher rates than rural enterprises and those led by women or older entrepreneurs.

Figure 7.3 Digital tool adoption rates among African SMEs.

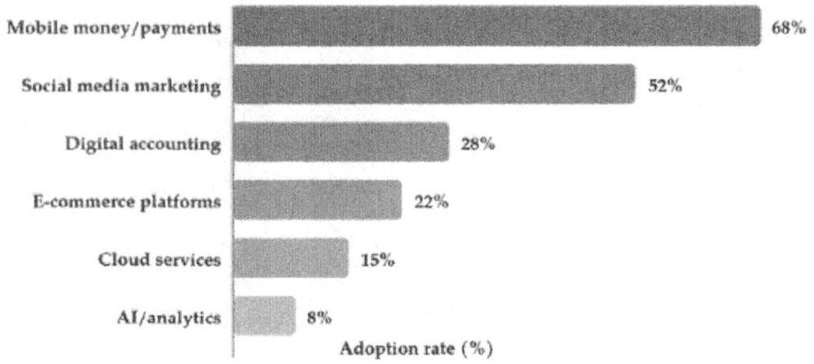

Source: IFC (2023).

Infrastructure issues—such as unreliable electricity, high data tariffs, and limited broadband—widen these gaps. Furthermore, the skills needed to use digital technologies effectively are scarce, especially in informal-sector businesses that employ most of Africa's workforce.

Box 7.2. Comparative Insight: SME Digitization in Southeast Asia and Latin America

Indonesia's government-led program to digitize 30 million micro and small enterprises by 2024 combined subsidized Internet access, simplified e-commerce onboarding, and digital literacy training. Mexico's INADEM program provided matching grants for SME technology adoption, reaching over 200,000 enterprises. In Colombia, the iNNpulsa program combines technology training with access to finance and mentorship. These models illustrate the importance of integrated support—combining connectivity, tools, skills, and finance—rather than technology-only interventions.

Innovation Hubs, Incubators, and Accelerators

Over the past decade, Africa's technology innovation ecosystem has seen remarkable growth. The number of active innovation hubs—including incubators, accelerators, makerspaces, and co-working spaces—has increased from around 50 in 2010 to over 1,100 in 2024 (AfriLabs, 2024). These hubs offer essential infrastructure such as reliable electricity, high-speed Internet, and meeting spaces, along with business development services like mentorship, training, networking opportunities, and access to funding.

Leading innovation hub ecosystems include Kenya's Silicon Savannah, anchored by Nairobi's iHub, founded in 2010, and Nigeria's Yaba technology cluster in Lagos, home to companies such as Flutterwave, Paystack, and Andela, and sometimes called "Yabacon Valley." The tech scenes in Cape Town and Johannesburg, South Africa, are also growing. Emerging hubs are developing in Egypt (Cairo), Rwanda (Kigali Innovation City), Ghana (Accra Digital Centre), Morocco, Senegal (CTIC Dakar), and Tunisia (el Gazala).

Figure 7.4. Growth of Technology and Innovation Hubs Across Africa (2010–2024)

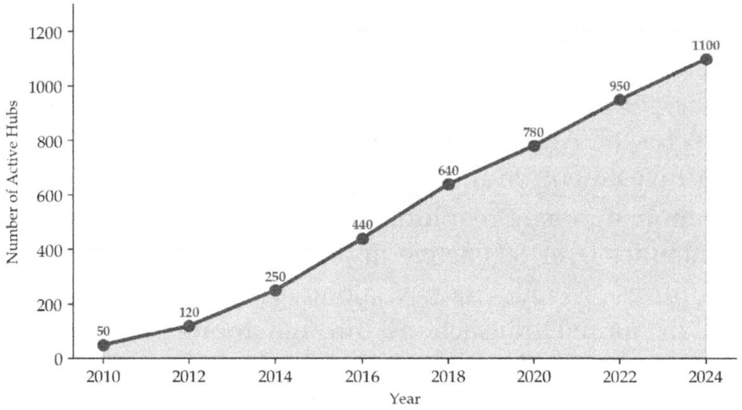

Sources: AfriLabs (2024); GSMA (2023).

Despite widespread interest in innovation hubs, ICT villages, and innovation cities across Africa, sustained success in creating thriving clusters of digital technology enterprises has primarily occurred in a few countries, such as Egypt, Mauritius, Morocco, Kenya, Tunisia, and South Africa. The smart village and Maadi Technopark in Egypt, Konza in Kenya, the Ebene cyber city in Mauritius, a series of tech parks in Morocco, and El Gahzala in Tunis, along with Gauteng and Cape Town in South Africa, are among the most successful examples. Cabo Verde, Ethiopia, Ghana, Senegal, and Tanzania are other countries that have launched digital innovation park initiatives.

The experience of African technology hubs shows that university-industry linkages, which are crucial for fostering indigenous innovation capacity (AfriLabs, 2024), remain underdeveloped. While top African institutions like the University of Cape Town, the University of Nairobi, Carnegie Mellon University Africa (Kigali), Makarere University, and Ashesi University (Ghana) are actively pursuing links between academic research and commercial applications, most African universities lack the funding, faculty incentives, and organizational structures needed to support technology transfer and the creation of spin-off companies effectively. Improving these linkages is therefore essential.

Fintech: Africa's Flagship Technology Sector

Financial technology is the most advanced and well-funded tech sector in Africa, capturing about 40–45% of all venture capital investment on the continent. Built on the success of M-Pesa's mobile money revolution in Kenya since 2007, the sector now includes over 800 active fintech startups across more than 40 countries. Mobile money alone processes over USD 900 billion in annual transactions through more than 300 million active accounts, making sub-Saharan Africa the global leader in mobile money adoption (GSMA, 2024).

The fintech landscape is diverse, spanning multiple segments including digital payments and transfers (Flutterwave, Paystack, Chipper Cash), digital lending and credit (Branch, FairMoney, Tala), digital insurance (InsurTech platforms offering micro-insurance for health, agriculture, and assets), savings and investment platforms (such as PiggyVest in Nigeria and M-Shwari in Kenya), remittances and blockchain-based financial services.

Table 7.2: Africa's leading start-up ecosystems by country

Country	Active Start-ups	VC Funding	Hubs	Key Cities	Leading Sectors
Nigeria	481	$1.7B (2023)	180+	Lagos, Abuja	Fintech, e-commerce, logistics, healthtech
Kenya	320	$800M (2023)	200+	Nairobi	Fintech, agritech, mobility, edtech
South Africa	290	$650M (2023)	150+	Cape Town, Johannesburg	Fintech, cleantech, enterprise software
Egypt	220	$520M (2023)	100+	Cairo	Fintech, logistics, SaaS, edtech
Ghana	95	$180M (2023)	60+	Accra	Fintech, healthtech, agritech
Rwanda	65	$110M (2023)	40+	Kigali	Fintech, healthtech, smart city solutions
Senegal	55	$85M (2023)	35+	Dakar	Fintech, agritech, e-commerce
Tanzania	50	$60M (2023)	30+	Dar es Salaam	Fintech, agritech, energy

Sources: Partech Africa (2024); AfriLabs (2024); Briter Bridges (2024).

Despite the dynamism in fintech, challenges remain in scaling enterprise activities at the hubs. Regulatory fragmentation across African countries complicates compliance and hampers cross-border expansion. Gaps in consumer protection, especially concerning predatory digital lending, require attention. Furthermore, financial inclusion is still inconsistent. While mobile money has successfully reached urban and semi-urban areas, the poorest and most remote communities, particularly women, are still excluded. Venture capital funding is heavily concentrated, with Nigeria, Kenya, South Africa, and Egypt receiving over 70% of investments, which limits ecosystem growth in other countries (GSMA, 2024).

7.4 Digital Trade, Platforms, and Services

The growth of digital trade and platform economies represents one of the most important economic transformations underway across Africa. E-commerce, digital payment platforms, business process outsourcing, digital content and creative industries, and cross-border digital trade facilitated by the African Continental Free Trade Area are reshaping how goods and services are produced, exchanged, and consumed (UNCTAD, 2023).

The AfCFTA provides a framework for expanding digital economic activities across borders. Figure 7.5 shows the estimated market value of Africa's digital economy in 2023, with projections for 2030, both measured in billions of USD.

The data show strong growth potential across all sectors, with some, like EdTech and AgriTech, expected to expand from a lower base, reflecting the continent's accelerating digital transformation. These figures highlight the potential for rapid growth in Africa's digital industries, with fintech and e-commerce leading the way. Services, platforms, and creative digital markets are projected to become high-growth areas by 2030.

Figure 7.5: Africa's digital economy by sector—estimated 2023 values and 2030 projections

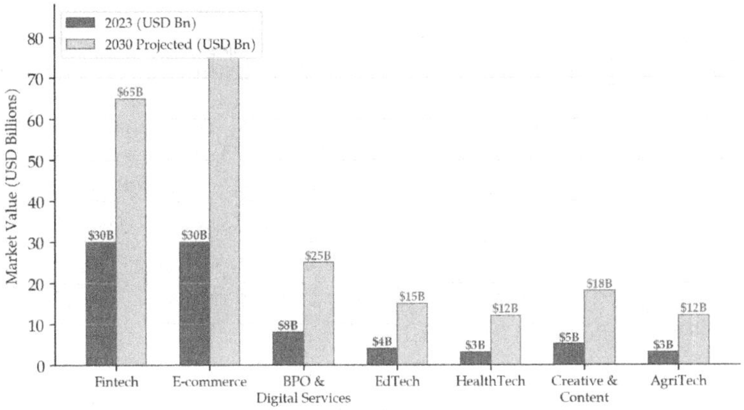

Sources: UNCTAD (2023); IFC (2023); Authors' estimates.

E-commerce and Digital Marketplaces

Africa's e-commerce market has grown substantially but remains small on a global scale. In 2023, its value was about USD 30 billion (roughly 3% of total retail sales), compared to online retail penetration rates of 25–30% in China and 15–20% in the United States.

Several factors impede e-commerce expansion, including high logistics costs—last-mile delivery in African cities can be 3–5 times higher than the global average due to poor addressing systems, traffic congestion, and infrastructure issues—low banking and card usage (partly offset by mobile money and cash on delivery), trust concerns (regarding product quality, delivery reliability, and returns), and regulatory gaps (such as cross-border e-commerce taxes, consumer protection, and data governance). The AfCFTA's digital trade protocol aims to address some of these challenges through standardized rules on electronic transactions, consumer protection, and data management.

Box 7.3 Comparative Insight: E-commerce Growth in Asia

Fueled by mobile-first platforms, integrated payment and logistics systems, and significant venture capital investments, Southeast Asia's e-commerce sector experienced rapid expansion, growing from USD 5 billion in 2015 to over USD 130 billion in 2023. While Shopee, Lazada, and Tokopedia dominate the region, domestic platforms like Indonesia's Bukalapak and Vietnam's Tiki have demonstrated the ability to thrive by adapting to local consumer preferences, payment methods, and logistical challenges. Asian examples underscore the importance of integrated ecosystems that seamlessly connect payments, logistics, and marketplaces, rather than treating them as separate services.

Business Process Outsourcing and Digital Services

Business process outsourcing (BPO) and digitally delivered services offer a major opportunity for African economies to participate in global services trade. The global BPO market exceeds USD 300 billion, but Africa's share remains modest at around USD 8 billion in 2023, though it is expanding.

Key sectors include customer service and call centers, with South Africa, Kenya, Ghana, and Morocco leading as destinations; South Africa alone employs over 270,000 BPO workers. Software development and IT outsourcing are other important areas, with companies like Andela, Gebeya, and Decagon training and placing African developers with global clients. Additionally, data annotation and AI training are rapidly growing segments where African workers label data for global AI systems, as exemplified by firms such as Samasource/ Sama in Kenya and Uganda.

Table 7.3 Africa's digital trade and platform economy—sectors, scale, and growth trajectories.

Sector	Key Platforms	Market Value	Growth Trajectory	Key Challenges
E-commerce (B2C)	Jumia, Takealot, Kilimall, Konga	$30B (2023)	Growing at 15–20% annually	Logistics costs, trust, and payment infrastructure
B2B Platforms	TradeDepot, Wasoko, Twiga Foods	$12B (2023)	Rapid growth, VC-backed	Working capital, supply chain complexity
Fintech & Payments	Flutterwave, Paystack, M-Pesa	$40B+ (2023)	40–45% of VC investment	Regulation, interoperability, consumer protection
BPO & IT Services	SA call centres, Andela, Sama	$8B (2023)	Steady growth, 10–15%/yr	Skills, infrastructure, and competition from Asia
Creative & Content	Nollywood, Boomplay, gaming studios	$5B+ (2023)	Rapid digital-driven growth	IP protection, monetisation, distribution
Gig & Platform Work	Uber, Bolt, Upwork, local platforms	$12B (2023)	25–30% annual growth	Worker protection, regulation, and income stability

Sources: UNCTAD (2023); IFC (2023); GSMA (2024).

Africa's digital content and creative sectors are also growing rapidly. Nollywood, Nigeria's film industry, generates an estimated USD 6.4 billion annually and employs about 1 million people. Driven by streaming services like Spotify, Apple Music, and the African platform Boomplay (with over 85 million users), Africa's music industry is expanding to international audiences.

Although still in its early stages, the gaming industry is expanding quickly, with studios in Nigeria, Kenya, South Africa, and Ghana creating content for both African and international markets. Furthermore, digital art, including the non-fungible token (NFT) marketplace, has opened new income streams for African visual artists and illustrators (UNCTAD, 2023).

7.5 Jobs, Work, and the Future of Skills

The impact of emerging technologies on employment is a critical and contested issue in Africa. The continent faces a significant job creation challenge, with an expected increase of 450 million in its working-age population from 2020 to 2035. This growth requires creating 18–20 million new jobs each year to accommodate new entrants into the labor force and reduce underemployment. At the same time, technology-driven automation threatens existing jobs, particularly in manufacturing, agriculture, and routine service sectors (International Labour Organization, 2024; WEF, 2023).

Job Creation Potential

Technology supports employment through various channels. The direct technology sector in Africa is projected to create 4–5 million jobs in 2024, mainly in software development, IT services, telecommunications, and digital platforms. The platform and gig economies—covering ride-sharing drivers, delivery couriers, freelance developers, and data annotators—are expanding rapidly, with estimates ranging from 5 to 15 million workers across the continent.

Technology-driven entrepreneurship in fields such as AgTech, HealthTech, EdTech, and FinTech creates new jobs. Also, emerging roles such as data scientists, AI specialists, cloud architects, cybersecurity analysts, and digital marketing professionals are growing, although current numbers are still small (ILO, 2024).

The most significant job creation often occurs indirectly. When traditional firms adopt new technology, they can increase productivity and competitiveness, driving expansion and more hiring. Moreover, digital platforms that connect informal-sector workers with customers (such as motorcycle taxi drivers and freelance plumbers) help formalize and grow existing employment rather than create entirely new jobs.

Job Displacement Risks

Automation and AI threaten jobs that involve routine, predictable tasks. The most at-risk roles include manufacturing assembly workers (as robotics and automation become viable even in lower-wage settings), agricultural laborers (due to advances in mechanization and precision agriculture), retail and wholesale clerks (with growing e-commerce and automated inventory systems), bank tellers and routine financial service workers (as digital banking expands), and data entry and basic administrative roles (becoming more automated through software tools and AI).

However, the pace and extent of job displacement in Africa are likely to differ from those in high-income countries for several reasons.

- Lower wage levels diminish the economic incentive for automation, leading to a longer payback period for replacing low-wage workers with robots compared to replacing high-wage workers.

- The informal sector, which employs most of Africa's workforce, largely operates outside the scope of formal automation.

- Infrastructure limitations, such as unreliable power and limited connectivity, further slow down technology adoption.

- Labor-intensive sectors such as agriculture and services may experience task transformation (changes in job duties) rather than complete job elimination (WEF, 2023).

Box 7.4 Comparative Insight: Automation and Employment in Asia

East Asia's experience provides both cautionary tales and encouraging examples. While China's manufacturing sector saw over 16 million jobs lost to automation between 2013 and 2020, the country simultaneously created millions of new positions in logistics, e-commerce, and digital services. Similarly, Vietnam's garment industry has adopted automation for tasks like cutting and quality inspection, yet overall employment has expanded due to productivity-driven export growth. In India, the IT services sector, which employs 5 million workers, faces potential disruption from AI-driven automation of routine coding and testing. Major Indian firms are investing heavily in reskilling programs. The key takeaway is that the net employment effect hinges on whether productivity gains lead to market expansion and new hiring opportunities.

The Evolving Skills Landscape

The skills needed in Africa's labor markets are changing quickly. Technical and digital skills, such as programming, data analysis,

AI and machine learning, cybersecurity, and cloud computing, are in high demand across many sectors, not just in tech. Cognitive and analytical skills—such as critical thinking, problem-solving, creativity, and decision-making—become increasingly important as automation takes over routine tasks. Socio-emotional skills, including communication, teamwork, adaptability, and emotional intelligence, are crucial for the service-focused, team-based work environments that make up much of the modern economy. Although the shift is slow, demand for manual and routine skills is dropping as automation grows (WEF, 2023).

Figure 7.6 illustrates the projected shifts in skill demands in Africa's labor market from 2020 to 2030, highlighting a move toward higher-level skills. The chart shows a consistent, sharp increase in demand for technical and digital skills, rising from about 20% in 2020 to nearly 48% by 2030. This trend reflects rapid digital transformation, automation, and the expansion of data-driven industries.

Figure 7.6. Projected Evolution of Skills Demand in Africa's Labour Market (2020–2030)

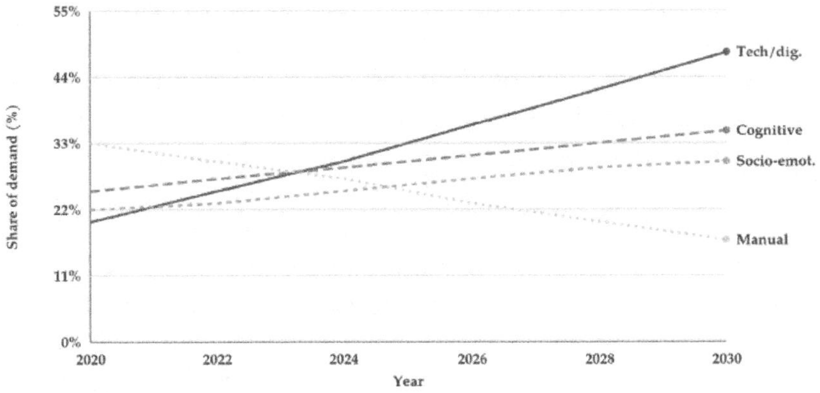

Sources: WEF (2023); ILO (2024).

Cognitive and analytical skills are also expected to grow steadily, highlighting the growing importance of problem-solving, critical thinking, and complex decision-making. Similarly, socio-emotional skills—including communication, adaptability, and collaboration—are projected to rise, indicating a growing appreciation for interpersonal and human-centered abilities in changing work environments influenced by emerging technologies. Meanwhile, demand for manual and routine tasks is expected to decline over the same period.

Addressing this skills transition requires a comprehensive approach: education systems must be reformed to focus on computational thinking, problem-solving, and digital literacy starting from primary school; technical and vocational training should be expanded and aligned with industry needs; coding bootcamps, micro-credentials, and other fast-track skills programs need to be scaled up; investments in lifelong learning and reskilling initiatives are essential for workers displaced by or at risk of automation; and the private sector should participate in skills development through apprenticeships, co-designed curricula, and on-the-job training (ILO, 2024).

Gender and Youth Dimensions

The gender gap in technology and employment remains a significant issue in Africa. Across sub-Saharan Africa, women are underrepresented in tech entrepreneurship, with women-led startups receiving about 15% of funding. Women are also underrepresented in STEM education and in the technology sector. Without targeted, gender-responsive strategies—such as digital literacy programs, support for women's entrepreneurship, and policies to address barriers like unpaid care work and gender discrimination—the digital economy risks widening existing gender inequalities (GSMA, 2024).

Africa's large youth population presents both a significant opportunity and a major policy challenge. With a median age of about 19 years and 60% of the population under 25, the continent has the youngest workforce in the world. However, youth unemployment and underemployment rates remain high—exceeding 30% in many countries, with young women disproportionately affected.

Figure 7.7: Gender gaps in digital economy participation (left) and youth employment status (right) in Africa

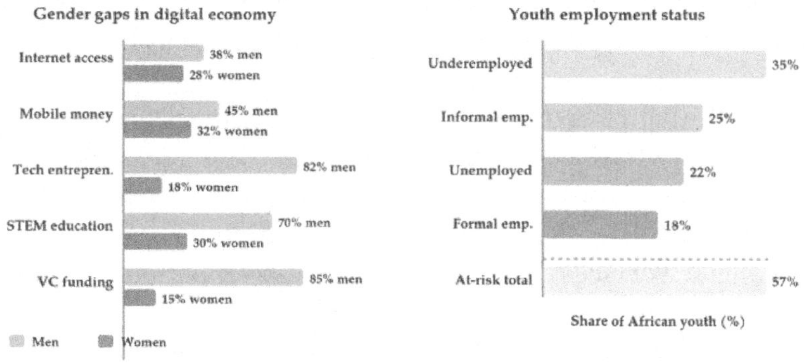

Sources: IFC (2023); ILO (2024); GSMA (2024)

To maximize the benefits of this demographic dividend, efforts should focus on increasing access to quality education and skills training, creating an environment that supports youth entrepreneurship, developing labor market intermediation services (connecting young people with opportunities), and ensuring that technology-driven economic growth creates jobs for young people, not just a highly educated elite (ILO, 2024).

Social Protection in the Digital Economy

The rise of platform and gig work raises important questions about social protection for workers outside traditional employment. Platform workers—including ride-sharing drivers, delivery couriers, freelance developers, and data annotators—often lack access to essential benefits such as health insurance, pension contributions, unemployment benefits, and workplace protections. Potential solutions being explored and tested include portable benefits systems (where benefits are tied to the worker rather than the employer), platform contribution mandates (requiring platforms to contribute to workers' social insurance), universal basic income experiments, and cooperative platform models (worker-owned alternatives to corporate platforms). Emerging regulatory responses in Africa include South Africa's Employment Services Amendment Bill and Kenya's Digital Economy Taxation proposals (ILO, 2024).

Overall, the assessment of technology's impact on employment indicates that manufacturing faces the greatest risk of displacement but also creates new roles, such as robotics technicians, quality engineers, and digital twin specialists. Financial services and agriculture also experience significant exposure to automation. Yet, both sectors are transitioning toward higher-value digital roles—such as AI specialists, cybersecurity experts, drone operators, and AgTech entrepreneurs—reflecting structural change rather than just job loss.

Conversely, the health, creative, and digital sectors face lower automation risks because of the ongoing need for distinctly human skills. In healthcare, technology mainly assists professionals, boosting demand for roles in health informatics, telemedicine coordination, and AI-driven diagnostics. The transport and logistics sector finds a middle ground, with gradual automation in fleet management and route optimization, while labor-intensive last-mile delivery remains a major source of employment.

Table 7.4: Technology's impact on employment by sector

Sector	Displacement Risk	New Roles Emerging	Key Skills Needed	Transition Dynamics
Manufacturing	35% tasks automatable	Robotics technicians, quality engineers, digital twins	Machine operation, maintenance, and industrial IoT	Transition from assembly to technology-augmented roles
Agriculture	25% tasks automatable	Drone operators, data analysts, AgTech entrepreneurs	Precision agriculture, farm management software	Mechanisation combined with knowledge-intensive roles
Financial Services	28% tasks automatable	AI specialists, cybersecurity, UX designers	Data science, regulatory technology, customer experience	From branch-based to digital-first services
Health	10% tasks automatable	Health informatics, telemedicine coordinators, AI techs	Digital health, genomics, medical device engineering	Technology augments rather than replaces health workers
Creative/ Digital	5% tasks automatable	Content creators, game developers, digital marketers	AI tools, platform management, data analytics	Technology-enabled expansion of creative employment
Transport/ Logistics	22% tasks automatable	Fleet managers, logistics analysts, drone operators	Route optimisation, autonomous vehicles, IoT	Gradual automation; last-mile remains labour-intensive

Sources: ILO (2024); WEF (2023).

This uneven impact of technological change on sector employment underscores the need for targeted reskilling strategies tailored to each industry's specific transition.

7.7 Conclusion

Emerging technologies and their economic applications are transforming Africa's industrial sector, entrepreneurial environment, trade patterns, and job markets. Industry 4.0 technologies provide opportunities to improve manufacturing and enable African businesses to participate in global supply chains. Digital tools are increasing productivity, expanding market access, and accelerating financing for millions of small and medium-sized enterprises. Platform economies are creating new forms of commerce and service delivery. Additionally, a vibrant startup ecosystem, primarily led by fintech but growing more diverse, is generating innovative solutions to African development challenges.

However, the distribution of these benefits remains unequal. The technology gap between leading economies (Nigeria, Kenya, South Africa, Egypt) and the rest of the continent, between formal and informal sectors, between urban and rural areas, between men and women, and between a small digital elite and the broader population threatens the inclusive growth envisioned by Africa's Agenda 2063.

Addressing this disparity requires intentional policy actions across multiple areas: investing in digital infrastructure and energy access as essential foundations; building human capital through reforming education and expanding skills training; developing flexible regulatory frameworks that encourage innovation while safeguarding workers and consumers; mobilizing diverse sources of financing—from venture capital to development finance to innovative tools like green bonds and diaspora investment; and ensuring that the AfCFTA's digital trade protocols establish a truly continental market that benefits smaller enterprises as well as larger ones.

Experiences from Asia, Europe and Latin America show that technology-driven economic transformation is neither automatic nor inherently inclusive. Ensuring that the benefits of technological progress are widely shared requires proactive, evidence-based stewardship. This includes combining industrial policies, education reform, social protection measures, and international cooperation.

References

African Development Bank. (2024). African economic outlook 2024: Driving Africa's transformation—the reform of the global financial architecture. AfDB.

African Union & UNIDO. (2022). Industrialise Africa: Strategies, policies, institutions, and financing. African Union Commission.

AfriLabs. (2024). Building a thriving innovation ecosystem: The state of Africa's tech hubs 2024. AfriLabs.

Briter Bridges. (2024). Africa's innovation economy: Annual funding report 2024. Briter Bridges.

GSMA. (2024). State of the industry report on mobile money 2024. GSMA.

International Finance Corporation. (2023). Investing in African SMEs: The role of digital tools and technologies. IFC.

International Labour Organization. (2024). World Employment and Social Outlook: Trends 2024. ILO.

Partech Africa. (2024). Africa Tech Venture Capital Report 2024. Partech Partners.

United Nations Conference on Trade and Development. (2023). Digital economy report 2023: Africa. UNCTAD.

United Nations Industrial Development Organization. (2022). Industrial development report 2022: The future of industrialisation in a post-pandemic world. UNIDO.

World Economic Forum. (2023). The future of jobs report 2023. WEF.

PART IV

Building Foundations, Seizing Opportunities and Managing Risks

Chapter 8: Building Foundations for Adopting Emerging Technologies

8.1. Introduction

The successful adoption of emerging technologies in Africa primarily depends on the strength of the foundational elements—the infrastructure, institutions, human capital, and governance systems that support innovation, enable their deployment, and promote sustainable impact. Analysis consistently shows these foundations as the main constraints on Africa's technology-driven transformation: infrastructure quality accounts for 40–60% of the variation in technology adoption rates among African countries, while governance quality accounts for an additional 35–50% (World Bank, 2023; International Telecommunication Union, 2024).

This chapter explores five key foundational areas: digital and physical infrastructure (Section 8.2), energy systems and reliability (Section 8.3), digital public infrastructure (Section 8.4), Financing (Section 8.5), human capital and innovation systems (Section 8.6), and institutions, regulation, and governance (Section 8.7). For each area, the analysis evaluates current conditions, identifies gaps, highlights progress across the continent, and incorporates comparative insights from Asia, Europe, Latin America, and, where relevant, Canada and

Oceania (Australia and New Zealand). The chapter ends with an integrated assessment of multidimensional readiness profiles and a hierarchical framework for understanding how these foundational layers interact.

8.2 Digital Infrastructure

Infrastructure—both digital and physical—forms the essential foundation for adopting all emerging technologies. Without reliable connectivity, access to computing resources, and adequate physical infrastructure (such as roads, ports, and buildings for technology deployment), even the most promising innovations cannot scale. Africa's connectivity infrastructure has undergone a remarkable transformation over the past two decades, but significant gaps remain—especially in rural areas, landlocked countries, and the most fragile states.

Broadband and Mobile Connectivity

Mobile broadband reaches 84% of sub-Saharan Africa's population in 2024. However, actual usage lags—only 28% of people use mobile Internet regularly in 2024—highlighting the gap between coverage, availability, and use. There is significant variation between countries. Morocco had 96% 4G coverage and 88% Internet penetration, while Egypt had 94% 4G coverage and 72% Internet penetration. Kenya had 92% 4G coverage in 2024. Meanwhile, countries like the Democratic Republic of the Congo face substantial gaps, with just 35% 4G coverage and 23% Internet penetration (ITU, 2024).

ITU data further shows that only 47% of Africa's population had Internet access in 2024, compared to the global average of 67%. Within Africa, access and usage vary widely by region, income level, age, gender, and geography. Internet penetration ranges from 71% in North Africa—close to the average for middle-income countries—to just 33% in Central Africa, where political instability, limited investment, and challenging geography converge (ITU, 2024). This places Africa significantly behind Latin America and the Caribbean (75%) and Asia-Pacific (62%).

Figure 8.1. Internet Penetration Across African Sub-Regions Compared with Global Benchmarks (2024).

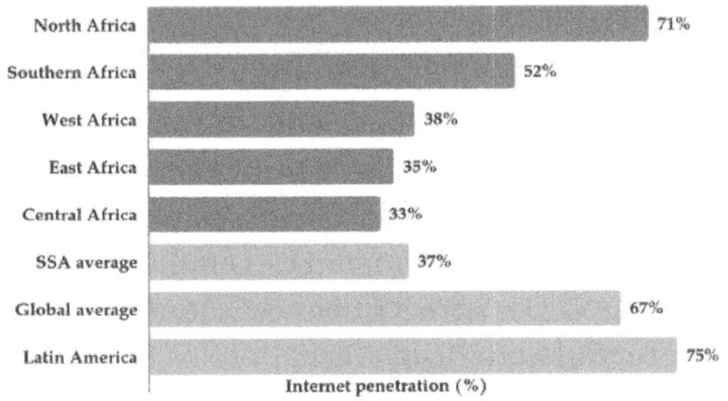

Source: ITU (2024); Alliance for Affordable Internet (2024).

Fifth-generation (5G) deployment remains concentrated in a few countries. By 2025, 29 African countries had commercial 5G services. The rollout of 5G in other countries faces challenges, including spectrum allocation costs, the need for extensive base-station infrastructure, limited revenue potential given income levels, and competition for public investment with more basic connectivity needs. The ITU (2024) predicts that 5G will reach only 20% of Africa's population by 2030, compared to 80% worldwide.

Low Earth Orbit satellite constellations—led by Starlink, OneWeb, and Amazon's Project Kuiper—provide an alternative for remote connectivity, with Starlink achieving download speeds over 100 Mbps in underserved areas of Nigeria, Rwanda, Kenya, and Malawi. However, equipment costs ($600) and monthly fees ($60+) remain prohibitively expensive for most users, limiting the market and usage to businesses, NGOs, and wealthy consumers (Starlink, 2024). The ITU projects that satellite technologies will serve 10–15% of Africa's rural connectivity needs by 2030, supplementing rather than replacing terrestrial networks.

Data affordability is a key factor affecting actual connectivity. The ITU Broadband Commission targets a goal where 1 GB of mobile data costs no more than 2% of Gross National Income (GNI) per capita. Research ICT Africa's RAMP Index data shows that by 2025, twelve countries had exceeded this threshold. In Madagascar, Malawi, Niger, the Central African Republic, and Chad, the cost exceeds 9%, making it nearly impossible for most people to afford Internet access. The Central African Republic, Democratic Republic of the Congo, Gabon, and South Sudan face particularly severe affordability issues.

Figure 8.2. Cost of 1 GB of Mobile Data as a Percentage of Monthly GNI per Capita (2024)

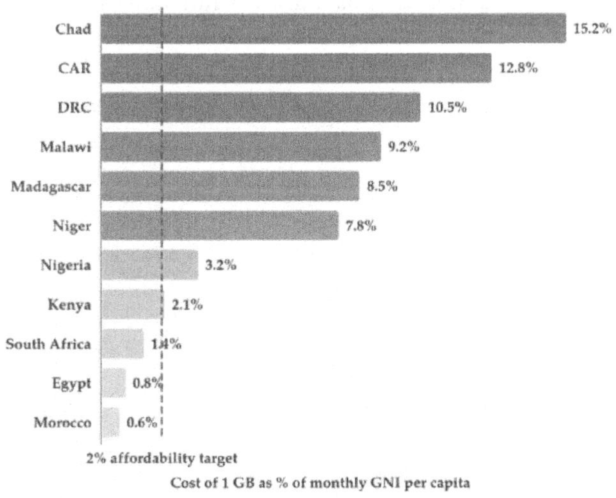

Source: Alliance for Affordable Internet (2024).

The European Union's Digital Compass target—1 GB of data costing less than 1% of the average monthly income by 2030—has already been achieved in most member states. India's aggressive spectrum auction strategy and managed competition have reduced the cost of 1 GB to below 0.9% of GNI per capita, serving as a model for increasing demand through pricing policies.

Submarine Cables and Backbone Networks

Submarine fiber-optic cables form the backbone of Africa's international connectivity. The continent has seen significant growth in submarine cable deployments: from having no operational submarine cables in 2000 to 18 systems by 2024, providing a total capacity of over 200 terabits per second (Tbps). In 2025, the continent's available design capacity increased by more than 600 Tbps, fueled by the launch of the 2Africa and Equiano cables. The East African coast has become a key investment hub, with Mombasa (Kenya) and Dar es Salaam (Tanzania) emerging as major cable landing points.

Figure 8.3. Growth of Submarine Cable Infrastructure Links to Africa

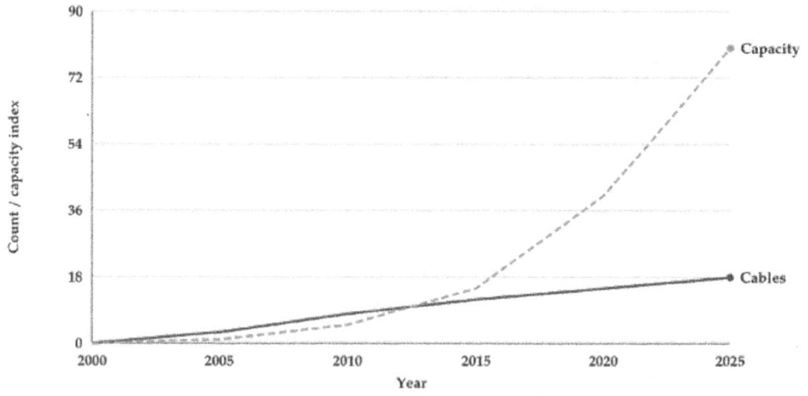

Source: TeleGeography (2024); ITU (2024).

In effect, submarine cable access provides a substantial boost for the adoption of emerging technologies. Countries with direct access to submarine cables experience bandwidth costs that are 40–60% lower than those of landlocked nations, underscoring the infrastructure's economic significance.

Terrestrial fiber networks extending from cable landing points into the interior of countries remain the main bottlenecks in most African countries. While countries such as Algeria, Cote d'Ivoire, Ghana, Kenya, Morocco, Nigeria, Senegal, South Africa, and Tunisia have built extensive national fiber backbones, many—particularly in Central and West Africa—lack terrestrial infrastructure to deliver international bandwidth to secondary cities and rural areas. Last-mile connectivity options, including fixed wireless access, Low Earth Orbit satellite networks like Starlink and OneWeb, and community networks, are emerging as alternatives to fiber. However, they remain expensive relative to local incomes.

Data Infrastructure

Data acts as the basic resource for almost all new technologies. AI systems require training data. IoT networks produce sensor data. Blockchain keeps transaction data, and big data analytics examines large datasets. Precision agriculture depends on soil, weather, and market data. Health tech uses patient and epidemiological data.

The data infrastructure layer covers the entire data value chain: generation (gathering data from diverse sources), storage (secure, accessible repositories), processing (computing to derive insights), and governance (rules for access, use, privacy, and sovereignty).

Africa faces significant data and data infrastructure challenges. Data generation is limited by low digitization of government services, health systems, and schools; minimal deployment of sensors and IoT; and gaps in surveys and statistical data, especially at sub-national levels. Data storage is scarce: *DataCenterMap*[1] lists over 11,000 public, private, and research data centers worldwide at the start of 2026, with 268 (2.5%) in

---

1 https://www.datacentermap.com/

Africa, even though the region accounts for 17% of the world's population. The distribution is highly uneven: South Africa, Nigeria, Kenya, and Egypt make up half of the continent's data centers. Furthermore, data centers face 40–60% higher total ownership costs than in developed markets, mainly due to unreliable power, higher cooling costs in hot climates, and a shortage of skilled workers (World Bank, 2024).

Compute Infrastructure

The compute infrastructure provides the processing power necessary to run applications, train AI models, host cloud services, and process IoT data streams. Compute infrastructure encompasses cloud computing (public, private, and hybrid), data centres (colocation and hyperscale), edge computing (processing at or near the data source), and high-performance computing for research. Global compute requirements are growing exponentially: training a large language model in 2023 required approximately 10,000 times as much compute as in 2018 (Epoch AI, 2023).

Africa's compute infrastructure is expanding rapidly, but from a low base. Major cloud providers (Amazon Web Services, Microsoft Azure, Google Cloud, and Huawei Cloud have established or announced presences in South Africa, Kenya, Nigeria, and Egypt, expanding access to cloud services. However, most African countries lack local cloud infrastructure, meaning data has to travel long distances, increasing latency and often being processed elsewhere, raising sovereignty concerns. Edge computing—processing data locally on devices or nearby servers—holds great potential for Africa by reducing reliance on centralized infrastructure and supporting applications in areas with limited connectivity.

Internet Exchange

Limited Internet exchange point infrastructure causes much intra-African Internet traffic to be routed internationally—sometimes traveling thousands of kilometers overseas before returning to destinations within the same city. The Internet Society (2024) reports that only about 50 Internet Exchange Points are active across Africa, compared to over 300 in Europe. Kenya, Nigeria, and South Africa lead in IXP numbers. The Kenyan IXP (KIXP) demonstrates the value of IXPs: it saves Kenyan users an estimated $20 million annually in bandwidth costs while improving speeds and reducing latency (Kenya ICT Authority, 2024). UNCTAD (2024) estimates that 85% of content accessed by African users is hosted outside the continent, creating ongoing demand for bandwidth and related costs. Concerns about data sovereignty—the need to store and process African data within African jurisdictions—are encouraging investment in local data centers, though economies of scale favor regional rather than national facilities (ITU, 2024).

Box 8.1 Comparative Insight: Connectivity Infrastructure in Asia, Latin America, and Australia

India's BharatNet programme, the world's largest rural broadband project, aims to connect 250,000 village panchayats with fibre-optic broadband and has completed connections to over 190,000 by 2024. Indonesia's Palapa Ring project connected the country's 514 districts with a 35,000 km fibre-optic backbone. In Latin America, Brazil's Norte Conectado programme deployed fibre-optic cables along 10,000 km of Amazon River routes to connect isolated communities. These large-scale government-led programmes demonstrate that deliberate public investment in last-mile connectivity can overcome market failures that leave remote populations unconnected.

8.3 Energy Systems and Reliability

Energy access and reliability are the biggest barriers to the adoption of emerging technologies across Africa. For technology deployment, energy is a crucial prerequisite that determines the success of all subsequent innovations (IEA, 2024). Energy is essential for powering data centers, cell towers, charging devices, and the entire digital economy. The International Energy Agency reports that about 600 million Africans—43% of the continent's population—lack electricity. At the same time, those with grid connections often experience outages that disrupt economic activities and weaken the reliability of digital infrastructure.

Sub-Saharan African companies face an average of 56 power outages per year, each lasting about 4.2 hours, resulting in over 235 hours of power loss annually (World Bank Enterprise Surveys, 2024). The African Development Bank (AfDB, 2024) estimates that outages cost Sub-Saharan businesses roughly 6% of their annual sales.

Regional disparities are significant. North Africa has nearly universal electricity access (98%) with fairly reliable grid systems. At the country level, Morocco and Egypt have achieved full access (100%), followed by South Africa and Ghana (85%), Kenya (75%), and Senegal (70%). Nigeria (62%), Ethiopia (51%), and Tanzania (40%) show moderate-to-low levels of access, while the Democratic Republic of the Congo (19%) has very limited national electricity coverage. Even where access exists, reliability remains an issue for technology-dependent activities: Nigeria experiences an average of 12.4 power outages per month, each lasting 6.2 hours, while the DRC experiences 15.2 outages per month, each lasting 8.5 hours. About 75% of Nigerian firms own backup generators, raising operating costs by 10–15% and effectively acting as a regressive tax on businesses (World Bank, 2023).

Figure 8.4. Electricity Access and Grid Reliability: African Regions vs. EU and Latin America

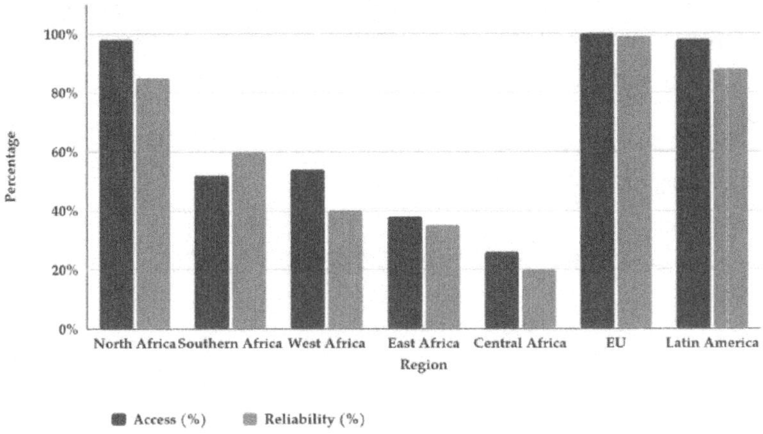

Source: IEA (2024); World Bank (2024).

The Renewable Energy Revolution

Africa has abundant renewable energy resources. The International Renewable Energy Agency estimates the continent's technical renewable energy potential at over 15,000 GW—about 500 times its current installed capacity. Solar photovoltaic (PV) deployment has grown significantly, with Morocco's Noor Ouarzazate complex (580 MW of concentrated solar power) and Egypt's Benban Solar Park (1.8 GW) demonstrating the large-scale viability of PV.

South Africa leads the continent with about 2,500 MW of installed solar capacity through its Renewable Energy Independent Power Producer Procurement Programme (REIPPPP), one of Africa's most successful public-private partnership models, which has attracted $14 billion in private investment for 5,000 MW of renewable capacity. East Africa is a leader in geothermal energy: Kenya produces roughly 40% of its electricity from geothermal sources, with 950 MW installed at the Olkaria complex in the Great Rift Valley (IRENA, 2024). Morocco has developed 3,700 MW of renewable capacity and aims to reach 52% renewable electricity by 2030. Morocco's success relies on regulatory certainty, a government-backed off-take guarantee, and currency-hedging tools for foreign investors.

Figure 8.5 Africa's Renewable Energy Landscape

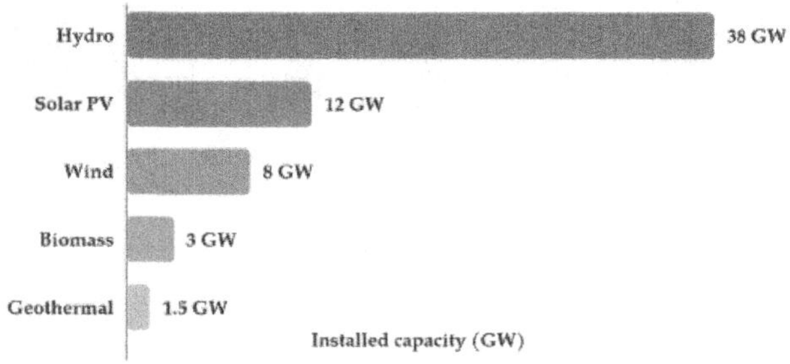

Source: IRENA (2024)

Overall, the continental renewable energy mix is evolving. Hydropower remains dominant, accounting for 48% of renewable generation, but its share is declining as solar (15%), wind (12%), biomass (17%), and geothermal (8%) expand. Wind energy has expanded significantly in South Africa (3,400 MW), Egypt (1,375 MW), and Morocco (1,200 MW), enabled by favourable wind resources along coastlines and elevated terrain.

Rural Electrification Pathways

Expanding access to electricity in rural and underserved areas is essential for increasing the adoption of emerging technology applications where they are needed the most. Rural electrification faces several challenges, including low population density and dispersed settlements, which significantly increase per-connection costs and make grid expansion less financially viable. Limited public funding, high perceived investment risks, aging transmission and distribution systems, affordability issues, regulatory uncertainties, land acquisition problems, limited local technical maintenance skills, and weak coordination between national and local authorities all slow down project implementation in rural areas.

Decentralized solutions such as mini-grids and solar home systems offer promising options for rural electrification. Solar home systems, enabled by pay-as-you-go (PAYG) business models pioneered by companies such as M-KOPA and BBOXX, now serve about 8 million households in East and West Africa.

Over 30 million African households now use solar home systems (World Bank, 2024), and more than 5,000 mini-grids are in operation (Africa Mini-Grid Developers Association, 2024). Customers make small daily or weekly payments via mobile money, making solar energy affordable for families earning as low as USD 2–5 per day. Mini-grids—small-scale electricity generation and distribution networks serving communities of 100–1,000 households—provide higher-quality electricity and can support productive activities like milling, welding, and cold storage. Scaling up requires clearer regulatory frameworks, standardized tariff structures, and capital (IEA, 2024).

Box 8.2 Comparative Insight: Energy Access in Asia, Latin America, Canada, and Oceania

India electrified its last unconnected village in 2018 through a combination of grid extension, solar microgrids, and solar home systems, supported by the Saubhagya scheme, which provided free connections to 26 million households. Bangladesh's Infrastructure Development Company Limited (IDCOL) facilitated the installation of over 6 million solar home systems—the world's largest such programme. In Latin America, Peru's national rural electrification programme combined grid extension with off-grid solar solutions to reach 98% electrification. These models demonstrate that universal access is achievable within a decade when supported by sustained political commitment, appropriate financing instruments, and technology-agnostic approaches that match solutions to local conditions.

Australia's experience with remote and Indigenous community electrification through stand-alone power systems (SAPS) and microgrids provides valuable lessons. The Australian Renewable Energy Agency (ARENA) has funded over 70 remote microgrid projects that integrate solar, battery storage, and diesel backup. New Zealand's electricity system, which derives over 80% of its generation from renewable sources, demonstrates the feasibility of high-renewables grids.

Canada's commitment to eliminating diesel dependency in remote communities by 2030 has spurred investment in community-owned renewable microgrids, with projects like the Gull Bay First Nation solar microgrid demonstrating how Indigenous-led energy transitions can combine local ownership, clean energy, and economic development—a model directly applicable to community-based electrification in Africa.

8.4 Digital Public Infrastructure

Digital Public Infrastructure (DPI)—foundational digital systems that enable identity verification, payments, and data exchange—is becoming increasingly vital for a country's ability to deliver digital services at scale. Built on broadband connectivity, data centers, and cloud environments, DPI includes digital identity systems (which enable individuals to prove their identities), fast payment infrastructure (for instant electronic transfers), and data exchange platforms (which facilitate secure information sharing across institutions). Countries that invest in these foundational layers can quickly and affordably deploy new digital services powered by emerging technologies; in contrast, those without them face repeated, costly, and fragmented efforts to build service-specific infrastructure (World Bank, 2023).

Assessment of DPI in Africa shows Rwanda at the top overall, with very high scores in digital identity and payment systems, although its data exchange framework is comparatively weaker. Kenya and Ghana also demonstrate strong payment system maturity—highlighting the continent's leadership in mobile money ecosystems. South Africa stands out for its high data exchange maturity, reflecting advanced interoperability and institutional digital integration.

Digital identity systems enable access to financial services, government benefits, health records, and commercial services. Rwanda's national identification system, integrated with the Irembo digital government platform, reaches 95% of adults and supports online access to more than 100 government services. Ghana's Ghana Card has registered over 16 million citizens (roughly 80% of adults), and research shows that cardholders use financial services 45% more than non-holders.

Figure 8.6. Digital Public Infrastructure Maturity Across Selected Countries in Africa

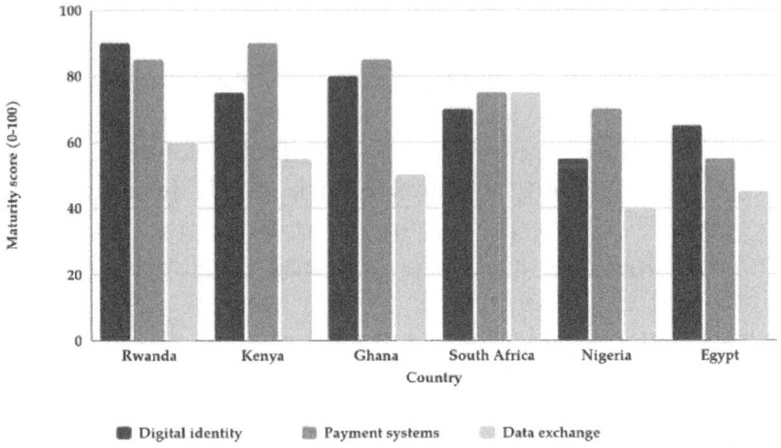

Source: World Bank (2023); ITU (2024); GSMA (2024).

Privacy and data protection concerns inevitably accompany the deployment of digital identity. Researchers and civil society groups advocate for privacy-by-design structures, independent oversight bodies, purpose-limitation safeguards, and robust legal frameworks governing the collection, storage, and use of identity data (African Union, 2022).

Africa's payment infrastructure has advanced more than any other component of the DPI, fueled by the mobile money revolution. The Pan-African Payment and Settlement System (PAPSS), launched by the African Export-Import Bank (Afreximbank) and the AfCFTA Secretariat, aims to facilitate instant cross-border payments in local currencies, reducing reliance on correspondent banking in Europe and North America. Overall, payment systems are the most developed DPI component; however, data exchange frameworks remain the weakest link.

Box 8.3 Comparative Insight: Digital Public Infrastructure —India's India Stack and Oceania

India's India Stack—which includes the Aadhaar biometric identity system (with 1.4 billion registrations), the Unified Payments Interface (UPI, handling over 10 billion monthly transactions), and the DigiLocker document exchange platform—is the most comprehensive DPI deployment worldwide. India Stack has enabled the government to transfer over USD 360 billion through direct benefit transfers, while reducing an estimated 25% of previously leaked funds. Australia's myGov digital identity and services platform and New Zealand's RealMe digital identity system provide additional models.

8.5 The Infrastructure Financing Gap

Financing remains the biggest challenge to Africa's infrastructure development. The AfDB estimates that the continent needs between $130 billion and $170 billion annually for infrastructure investment, but current investments amount to only $85–$90 billion annually, leaving a financing gap of approximately $88 billion (AfDB, 2024).

In Africa, domestic public resources account for about 45% of current infrastructure investment but are limited by weak tax collection (17% of GDP, compared to 34% in OECD countries) and high debt levels.

Private investment primarily focuses on telecommunications, while other sectors attract little private capital due to long payback periods, regulatory uncertainty, and political risks. Chinese infrastructure lending in Africa, which peaked at $30 billion annually in 2016–2017, has fallen to around $15 billion annually, mainly because of concerns about debt sustainability and shifting financial conditions (China Africa Research Initiative, 2024). Development Finance Institutions (DFIs) account for about 20% of infrastructure funding and play a catalytic role, especially in energy and multi-sector projects, by offering concessional finance, guarantees, and technical assistance to reduce investment risks.

Figure 8.7 Annual Infrastructure Investment: Needed vs. Current

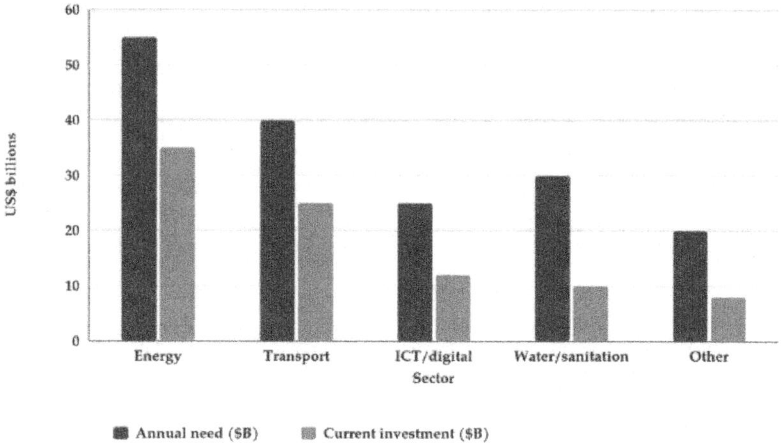

Source: AfDB (2024); ITU (2024); IEA (2024); UNCTAD (2024).

Innovative financing mechanisms are gaining traction. Blended finance instruments—combining concessional development finance with commercial capital—have proven effective in de-risking renewable energy projects in East Africa. Green bonds and sustainability-linked bonds attract climate-conscious investors. The African Development Bank's Room2Run initiative has securitized $1 billion in loans, freeing up balance sheet capacity for new infrastructure lending. Pension and sovereign wealth funds in Africa, managing over $2.5 trillion, remain largely untapped sources of patient capital for long-term infrastructure development (AfDB, 2024).

8.6 Human Capital, Research, and Innovation Systems

Human capital—the knowledge, skills, and abilities of a country's population—fundamentally influences how successfully emerging technologies are adopted. The World Bank's Human Capital Index ranks most African countries in the bottom quartile globally.

Africa's technological growth is hampered primarily by significant skills shortages. While the continent's young population provides a major demographic advantage, inadequate education systems, limited technical training opportunities, and ongoing brain drain hinder human capital development. UNESCO (2024) estimates that the continent needs to train 9 million more software developers by 2030 to meet projected demand, while it currently produces only about 200,000 technology graduates ready to work each year, resulting in an annual shortfall of 700,000.

Critical Skills Gaps Across Technology Domains

The demand for tech skills greatly exceeds supply across all domains in Africa, including cybersecurity, software development, DevOps, and AI. The World Economic Forum (WEF, 2024) reports that 68% of African tech employers struggle to find candidates with the right skills; when positions are filled, 52% of employers say new hires need 6–12 months of on-the-job training before becoming productive—much longer than the 1–3 months typical in developed markets.

The African Cybersecurity Centre (2024) estimates that African institutions will need 350,000 cybersecurity professionals, but only 23,000 are available—a 93% shortfall. Data science and artificial intelligence skills are also scarce: only 12 African universities offer graduate AI programs, producing fewer than

500 graduates each year compared to an industry demand of 6,000 (UNESCO, 2024).

Figure 8.8. Critical Skills Gap

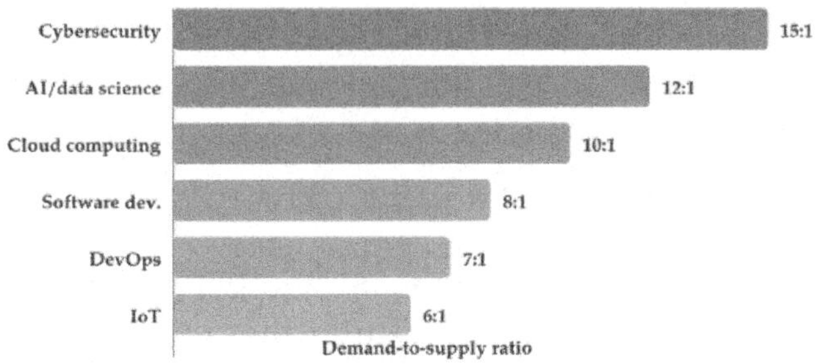

Source: World Bank (2023); UNESCO (2024).

Digital Literacy and STEM Education

Digital literacy remains limited across the continent. The ITU estimates that only 28% of Africans possess basic digital skills—defined as the ability to use a computer or mobile device for simple tasks, including email, web browsing, and file management. Gender and rural-urban gaps are significant: women are 23% less likely than men to have digital skills, and rural populations have literacy rates that are 45% lower than those in urban areas (GSMA, 2024).

Language is a subtle yet major barrier: most software interfaces and training materials are in English, French, or Arabic, creating access issues for speakers of hundreds of indigenous languages. Several countries are making efforts: Rwanda's Competency-Based Curriculum requires computer literacy from primary school; South Africa added ICT as a formal school subject in 2020; Kenya's Digital Literacy Programme distributed tablets to over 1.2 million primary school students.

At the tertiary level, STEM enrollment remains below global benchmarks: STEM fields account for about 22% of African university graduates, compared to 30% worldwide. Women comprise only 30% of STEM students. Africa invests just 0.4% of GDP in research and development, compared with global averages of 1.8% and the EU goal of 3%. Sub-Saharan Africa accounts for only 1.3% of global academic publications despite representing 14% of the world's population.

Figure 8.9. Human Capital and Research Capacity Assessment

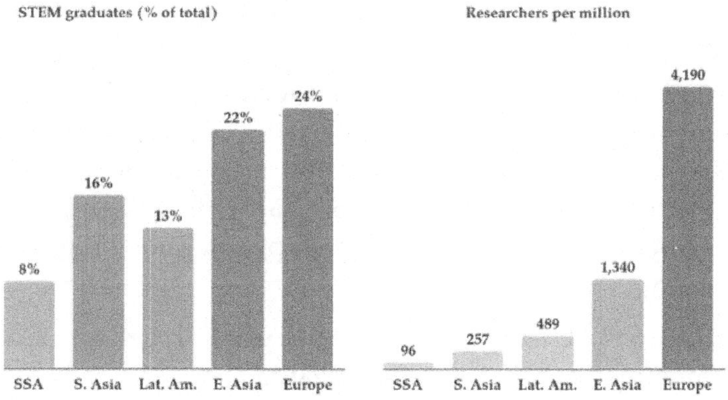

Source: World Bank (2023); UNESCO (2024).

Innovative training models outside traditional universities are helping to bridge the skills gap. Andela provides intensive software development training with job placement guarantees. Moringa School operates in Kenya and Rwanda, offering short-term software engineering bootcamps with an employment rate of over 80%. Kenya's Ajira Digital Program has trained 650,000 youth since 2016. Ethiopia's 5 Million Coder Initiative, South Africa's Presidential Commission on the Fourth Industrial Revolution, and Nigeria's 3 Million Technical Talent (3MTT) initiative are adopting similar strategies.

Brain Drain: Patterns, Drivers, and Impacts

Skilled migration significantly depletes Africa's human capital, particularly affecting technology professionals, as their skills are portable and international salary gaps are large. The UNDP (2024) estimates that around 2.3 million African professionals with higher education work outside the continent—18% of all African tertiary graduates—and this percentage rises to 35% among technology professionals.

Software engineers in Silicon Valley earn between $150,000 and $200,000, compared to $12,000 to $18,000 in Nigeria and $20,000 to $30,000 in South Africa for similar roles (PayScale, 2024). Europe attracts 42% of African technology migrants, North America 31%, and the Middle East 18%. Addressing brain drain requires improving local salaries and career opportunities, promoting circular migration, and leveraging diaspora networks for investment, mentorship, and knowledge transfer.

Box 8.4 Comparative Insight: Human Capital Development in Asia and Oceania

South Korea's transformation from one of the world's poorest countries (1960) to a global technology leader was built on sustained investment in education, rising from 2.6% to 5.1% of GDP over three decades. China's massive expansion of engineering education—producing over 4.7 million STEM graduates annually—was complemented by the return of diaspora scientists. Australia's Cooperative Research Centres (CRC) programme fosters collaboration between universities, industry, and government on applied research.

8.7 Institutions, Regulation, and Governance

Institutional quality significantly impacts technology adoption outcomes. Research indicates that governance quality—encompassing regulatory effectiveness, government capacity, rule of law, and accountability—accounts for 35–50% of the variation in technology adoption rates across countries (World Bank, 2023).

Mauritius ranks highest in both governance quality (77) and the E-Government Development Index (0.78), followed by South Africa (61; 0.74) and Rwanda (68; 0.65). Countries with a composite governance quality score above 65 have technology penetration rates that are 2–3 times higher than those with scores below 40.

Figure 8.10 Governance Quality and E-government Development Across Africa

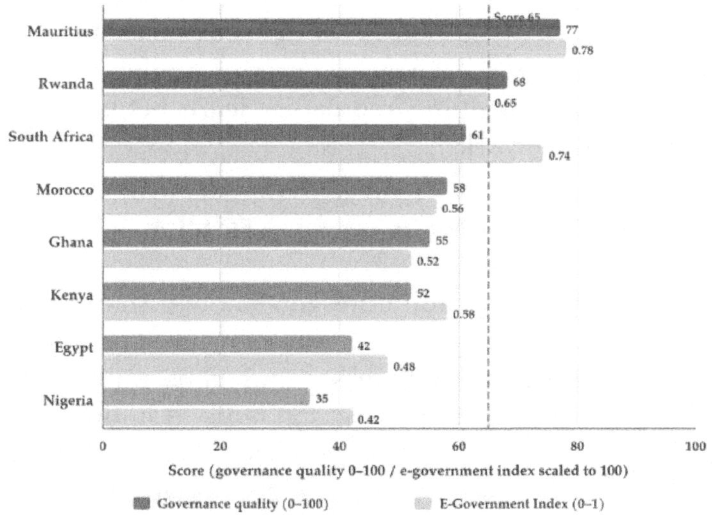

Source: World Bank (2023); United Nations (2024).

Regulatory Frameworks and Innovation

Effective regulation balances the promotion of innovation with the protection of consumers, workers, and the public interest. Kenya's Communications Authority exemplifies effective regulation: maintaining independence, allocating spectrum transparently, and promoting market competition. Kenya's light-touch approach to fintech regulation allowed M-Pesa's mobile money revolution to flourish before gradually increasing oversight.

UNCTAD (2024) identifies 847 distinct regulatory requirements across African countries that impact digital businesses. The World Economic Forum (2024) estimates that compliance costs for technology companies operating in multiple countries average $185,000 per country—a significant structural barrier to continental expansion.

Regulatory sandboxes—controlled environments where innovators can test new products with relaxed rules—have become vital tools for advancing emerging technologies. Rwanda's financial regulatory sandbox, Ghana's drone delivery sandbox, and Sierra Leone's fintech sandbox all promote innovation while helping regulators learn before establishing permanent frameworks.

Market Fragmentation and AfCFTA

Africa's political geography presents another significant barrier for technology adoption and commercialization. With many African countries having diverse regulations, multiple currency zones, and varying technical standards resulting in market fragmentation that limits economies of scale. The AfCFTA, operational since January 2021, offers opportunities to reduce market fragmentation. ASEAN's Digital Masterplan 2025 established a regional framework for interoperable digital payments and cross-border data flows across 10 countries, providing a practical model to decrease fragmentation.

8.8 Inequality, Inclusion and Gender Dimensions

Emerging technology adoption in Africa takes place amid significant inequality. Without deliberate policies to expand the benefits of technology to marginalized groups

Gender Disparities in Technology

Gender gaps affect all aspects of technology adoption in Africa. The GSMA (2024) reports that women are 15% less likely than men to own a mobile phone and 37% less likely to use mobile Internet. Women make up only 29% of technology-sector workers and 18% of tech entrepreneurs in Africa (UNESCO, 2024). In technical fields such as software development, data science, and cybersecurity, female representation often drops below 15%. These workforce gaps are driven by educational shortcomings, workplace challenges like harassment and discrimination, a lack of female role models and mentors, and cultural beliefs that view technology as a male domain.

Figure 8.11. Gender Participation Gaps Across Technology Dimensions in Africa

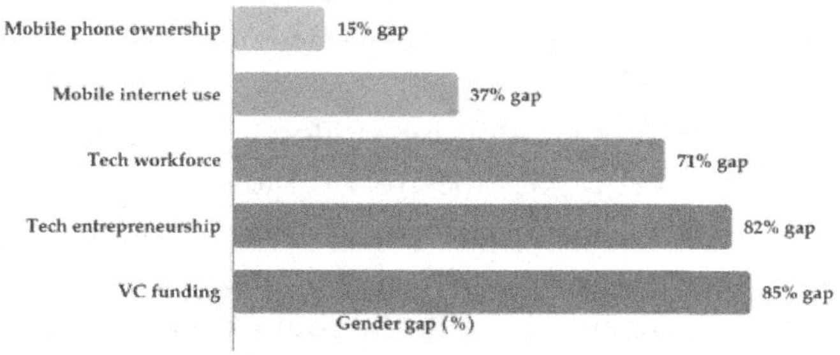

Source: GSMA (2024); UNESCO (2024); African Union (2024).

Youth Employment and the Demographic Dividend

Africa's youth population presents both a major opportunity and a significant risk (AfDB, 2024). Youth unemployment rates average 22%, more than twice the adult rate of 9%, while underemployment affects another 35% of young workers. Together, unemployment and underemployment affect 57% of African youth—leading to a substantial loss of human potential.

The technology sector offers some of Africa's fastest-growing job opportunities: the digital economy is expected to employ 100 million Africans by 2030, but accessing these opportunities requires skills that most youth currently lack. Targeted efforts—such as bootcamps, apprenticeships, online learning platforms with local-language content, and subsidized access to devices and data for students—are producing encouraging results.

8.9 Integrated Assessment

Understanding technology adoption requires evaluating the foundational areas discussed above simultaneously. Countries with balanced development across energy systems, DPI, human capital, governance, and broadband infrastructure—such as Egypt, Kenya, Mauritius, Morocco, and South Africa—also lead in the adoption of emerging technologies. In contrast, countries with the lowest levels—such as Burundi, the Central African Republic, Chad, and South Sudan—face major challenges.

Emerging technology adoption follows hierarchical patterns where higher-level capabilities rely on lower-level foundations. Level 1 (Physical Infrastructure) provides the basics: roads, electricity grids, and physical facilities. Level 2 (Digital connectivity, including broadband, data centres and compute) builds on physical infrastructure. Level 3 (Digital Services such as agriTech, eduTech, and FinTech) relies on connectivity. Level

4 (Human Capital comprising skills, talent and broad public digital literacy) develops and uses these platforms. Level 5 (Governance, including institutions, policies, and regulations) enables coordination across all lower levels.

Figure 8.12 Technology Adoption Foundation Pyramid

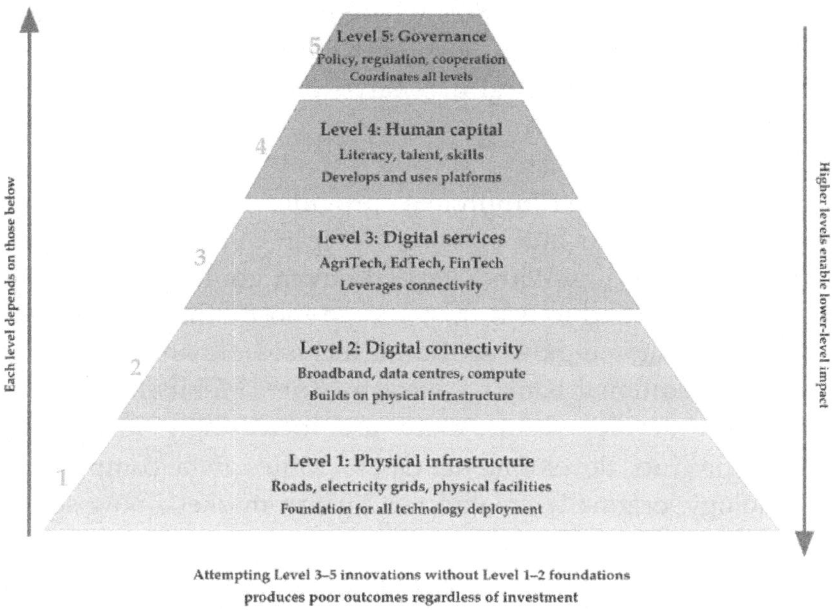

Source: Author's Synthesis

This hierarchy explains why technology initiatives frequently fail when foundations are inadequate: attempting Level 4 innovations (such as AI-powered services) without Level 2 connectivity or Level 1 electricity generates disappointing outcomes regardless of investment levels. The model also helps to identify opportunities for leapfrogging—where certain foundational layers can be partially skipped (as mobile money leapfrogged traditional banking infrastructure)—while noting that such opportunities are the exception rather than the rule.

8.10. Opportunity Areas for African Leadership

Despite major challenges, Africa possesses unique qualities that could enable it to lead in certain technology sectors. Instead of directly competing with established tech giants in their fields, Africa should focus on niches where its distinctive traits—such as demographic profile, resource availability, development environment, and cultural diversity—provide real competitive advantages.

Africa's mobile-first approach provides lessons for other developing regions and positions African companies to export innovation globally. With 73% of Kenyan adults using mobile money—compared to less than 10% in most OECD countries—Africa has shown how to build inclusive financial systems without traditional banking infrastructure. Mobile-first design principles developed in Africa are increasingly influencing global product development: Opera Mini's data compression technology, originally created for African markets, now serves 350 million users worldwide.

Necessity-driven innovation in resource-limited settings has produced uniquely African solutions with global relevance. Zipline's medical drone delivery system—designed for Rwanda's challenging terrain—has completed over 500,000 deliveries and now operates in the United States and Japan. mPharma's pharmacy management system reduces medicine costs by 40% through supply chain improvements across six countries.

Africa possesses about 30% of the world's mineral reserves, including 80% of platinum-group metals, 60% of cobalt, and large quantities of lithium, graphite, manganese, and rare-earth elements. The Democratic Republic of the Congo supplies 70%

of the world's cobalt. Global demand for lithium is expected to increase 40 times by 2040. However, Africa currently gains little from these resources: 80% of cobalt refining, 70% of lithium processing, and 90% of rare earth element processing take place in China. To benefit from these minerals, there needs to be a shift from exporting raw materials to adding value through processing.

Green hydrogen presents a highly promising opportunity, with potential exports surpassing $100 billion annually by 2050. Namibia's Hyphen Hydrogen Energy project plans to produce 3,000 MW of green hydrogen by 2030.

Africa's more than 2,000 languages present both a development challenge and an emerging AI opportunity. The Masakhane open-source research community has developed natural language processing tools for 27 African languages. South Africa's Lelapa AI develops large language models tailored for African languages and contexts. As global demand for multilingual AI increases—spurred by the EU's AI Act requirements for non-discrimination across languages—African expertise in low-resource language modeling could command a significant premium in the worldwide AI services market.

A comparison of opportunities shows that the largest gap lies in critical minerals, where the potential greatly exceeds current use, indicating that Africa's mineral resources for technologies such as batteries and electronics remain untapped. There is a significant gap between linguistic diversity and AI, highlighting the opportunity to develop AI systems that leverage Africa's rich linguistic diversity.

Figure 8.14. Africa's Opportunity Areas: Strategic Potential vs. Current Utilisation

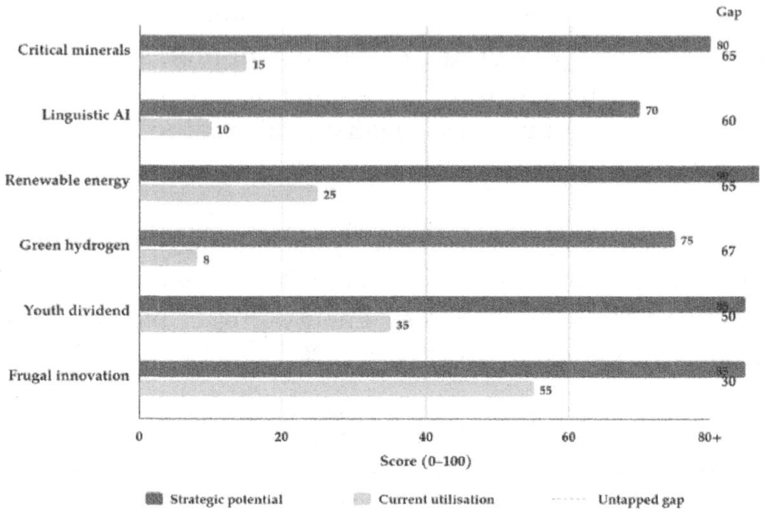

Source: McKinsey & Company (2024); IRENA (2024); GSMA (2024); USGS (2024).

Renewable energy also shows a substantial gap, suggesting that Africa's abundant solar, wind, and hydro resources are not yet fully harnessed. Youth dividend and frugal innovation show moderate gaps, reflecting strong demographic and innovation potential that is only partly tapped.

8.11 Conclusion

This chapter has examined the key elements that support successful technology adoption across Africa—digital and physical infrastructure, energy systems, digital public infrastructure, human capital and innovation systems, institutions, and governance—along with the challenges of skills shortages, market fragmentation, and inequality, as well as Africa's unique opportunity areas. Throughout these topics, the continent shows considerable diversity: countries like Mauritius, South Africa, Kenya, and Morocco demonstrate that building strong foundations is achievable, while many others face significant gaps.

Research consistently shows that foundations are the main drivers of the adoption and success of emerging technologies for development. Infrastructure quality, energy reliability, institutional effectiveness, and human capital account for more variation in technology adoption outcomes than the technologies themselves do. The infrastructure financing gap—$88 billion annually—is significant but not insurmountable through diversified strategies that combine domestic resource mobilization, blended finance, and innovative mechanisms. Skill shortages at crisis levels across all technology domains require both expanded traditional education and alternative training pathways.

Africa demonstrates remarkable innovation despite constraints. Mobile money adoption exceeds that of developed countries. Renewable energy resources position the continent for competitive and sustainable industrial growth. Critical mineral reserves provide strategic advantages in green technology supply chains. Linguistic diversity offers opportunities for AI development. The youth demographic dividend holds transformative potential if actively engaged. Unlocking Africa's technological potential requires addressing key challenges while strategically leveraging additional strengths. Experiences from Asia, the EU, Latin America, Canada, and Oceania demonstrate that targeted, well-funded foundation-building efforts can lead to transformative change within realistic timeframes.

References

Africa Mini-Grid Developers Association. (2024). Mini-grid market report: Africa. AMDA Publishing.

African Cybersecurity Centre. (2024). African cybersecurity workforce study. ACC Research Division.

African Development Bank. (2023). African Economic Outlook 2023. African Development Bank Group.

African Development Bank. (2024). African Economic Outlook 2024. AfDB Publishing.

African Union. (2022). African Union data policy framework. African Union Commission.

African Union. (2024). Agenda 2063 progress report. African Union Commission.

Alliance for Affordable Internet. (2024). 2024 affordability report. https://a4ai.org

Benchmark Mineral Intelligence. (2024). Battery raw materials outlook. BMI Research.

China Africa Research Initiative. (2024). Chinese loans to Africa database. Johns Hopkins SAIS.

European Commission. (2024). EU hydrogen strategy: Progress report. Publications Office of the EU.

GSMA. (2024). The mobile economy Africa 2024. GSM Association.

International Energy Agency. (2024). Africa Energy Outlook 2024. IEA Publications.

International Labour Organization. (2024). World Employment and Social Outlook: Africa. ILO Publishing.

International Renewable Energy Agency. (2024). Renewable energy market analysis: Africa. IRENA.

International Telecommunication Union. (2024). Measuring digital development: Facts and figures 2024. ITU.

Internet Society. (2024). Internet exchange points: Africa report. ISOC Technical Papers.

Kenya ICT Authority. (2024). Digital economy blueprint progress report. Government of Kenya.

Masakhane NLP. (2024). Participatory research for African language technology. https://www.masakhane.io
McKinsey & Company. (2024). Africa's technology ecosystem: State of play. McKinsey Global Institute.
Opera Software. (2024). State of the mobile web Africa 2024. Opera Software AS.
PayScale. (2024). Global salary comparison: Technology sector. PayScale Research.
Safaricom. (2024). M-PESA milestone report. Safaricom PLC.
Starlink. (2024). Africa expansion report. SpaceX Communications.
TeleGeography. (2024). Submarine cable map and database 2024. TeleGeography.
UNCTAD. (2024). Digital economy report: Africa edition. UNCTAD Publishing.
UNDP. (2024). Human development report 2024: Technology and inequality. UNDP Publishing.
UNESCO. (2024). Science report: Race against time for smarter development. UNESCO Institute for Statistics.
United Nations. (2024). UN e-government survey 2024. UN DESA.
USGS. (2024). Mineral commodity summaries 2024. USGS Publishing.
World Bank. (2023). World Development Report 2023. World Bank Group.
World Bank. (2024). World development indicators 2024. World Bank Publications.
World Bank Enterprise Surveys. (2024). Africa infrastructure constraints report. World Bank Group.
World Economic Forum. (2024). Global competitiveness report 2024: Africa regional analysis. WEF Publishing.
Zipline International. (2024). Global impact report: Drone delivery at scale. https://www.flyzipline.com

Chapter 9: Management of Risks and Governance of Emerging Technologies

9.1 Introduction

The adoption of emerging technologies in Africa offers both transformative opportunities, extensively discussed in this volume, and significant governance challenges and risks that require careful, proactive management. While earlier chapters addressed infrastructure limitations and strategic possibilities, this chapter highlights the importance of governance: maximizing the benefits of technology while minimizing potential harms through appropriate policies, regulations, institutions, and ethical frameworks rooted in African values and development goals.

In Africa, emerging technology governance faces significant challenges due to limited regulatory capacity, rapid technological advancements outpacing policy development, tensions between promoting innovation and managing risks, and the need to balance national sovereignty with regional and international cooperation.

Many emerging technologies—such as artificial intelligence, autonomous systems, blockchain, and robotics—raise new ethical questions without clear precedents, requiring Africa to help shape global governance standards rather than simply adopting frameworks created elsewhere. The stakes are high:

poor governance can increase inequality, violate human rights, enable excessive surveillance, and accelerate environmental damage, while effective governance can help technology support development goals, protect basic rights, promote inclusion, and position Africa as a responsible leader in technology.

This chapter discusses five main governance areas: ethical, social, and environmental risks (Section 9.2); data protection, privacy, and cybersecurity (Section 9.3); AI governance and algorithmic accountability (Section 9.4); standards, regulations, and regional harmonization (Section 9.5); and digital rights and Internet freedom within comprehensive governance strategies (Section 9.6).

9.2 Ethical, Social, and Environmental Risks

Emerging technologies present complex ethical, social, and environmental challenges that require proactive governance. These risks are interconnected—algorithmic bias has social consequences, surveillance technology raises ethical concerns, and digital infrastructure impacts the environment. Countries need to develop governance frameworks that address these diverse risks while promoting innovation and growth.

9.2.1 The Technology Risk Landscape in Africa

Technology risks in Africa differ from those in developed countries because of unique socioeconomic factors, weaker institutions, and limited governance capacity. The World Economic Forum (2024) highlights eight key technology risk areas impacting Africa, with varying levels of severity, likelihood, and preparedness.

What is immediately clear is the widespread lack of readiness across all categories: the average preparedness score of 36 sharply contrasts with the average severity of 84 and the likelihood of 82, indicating serious and urgent governance gaps.

Figure 9.1. Technology Risk Assessment for Africa: Severity, Likelihood, and Preparedness

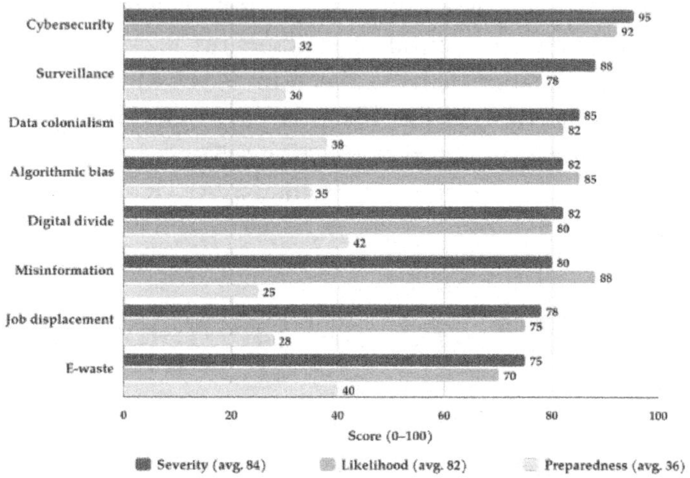

Source: African Union (2024); WEF (2024).

Cybersecurity threats are the most severe and likely among all risk categories, yet preparedness remains very low in Africa, creating serious vulnerabilities in an increasingly digital continent. Surveillance abuse also shows high severity with minimal preparedness, highlighting limited legal oversight on government surveillance. Misinformation has the lowest preparedness score (25), especially given the rapid increase in social media use and its documented role in fueling electoral violence and public health crises across Africa.

Job displacement caused by automation—though rated lower in severity than cybersecurity—receives the lowest preparedness score (28), indicating inadequate policy responses to structural shifts in employment. This is especially concerning given

Africa's youth unemployment crisis and the continent's need to create an estimated 20 million new jobs each year to support labor force growth (ILO, 2024). Addressing this risk requires substantial investment in regulatory institutions, technical capacity, legal frameworks, and multi-stakeholder engagement (African Union, 2024).

9.2.2 Ethical Frameworks for Technology: African Perspectives

Ubuntu and Communitarian Ethics

African ethical frameworks for technology should embody continental philosophical traditions, especially Ubuntu—the communitarian philosophy expressed in the Nguni phrase Umuntu ngumuntu ngabantu (a person is a person through other people). Ubuntu emphasizes collective benefits over individual gains, suggesting that technology should prioritize community development over personal advancement. This contrasts sharply with Western ethical frameworks, which focus on individual rights and autonomy.

Ubuntu-inspired technology governance highlights communal data ownership, considers access to technology as a collective right, and holds algorithmic systems accountable to the communities impacted by their decisions. The African Union (2024) argues that Ubuntu perspectives also shape views on data sovereignty: data about communities—such as health patterns, farming practices, and cultural knowledge—belongs collectively to communities rather than to individuals or corporations that extract it. This approach impacts consent processes, benefit-sharing arrangements, and governance frameworks for community data.

Several governance initiatives already incorporate Ubuntu principles. South Africa's National Policy Framework for

Artificial Intelligence explicitly highlights Ubuntu as a core value (South African Department of Communications, 2024). The African Declaration on Internet Rights and Freedoms references African community values alongside universal human rights. Kenya's Data Protection Act includes provisions for community data rights that extend beyond individual consent.

Besides Ubuntu, decolonial perspectives are gaining importance among think tanks, raising questions about whose values shape technology design, who benefits from its deployment, and who bears the risks. African scholars point out that much of what is called "universal" technology ethics reflects Western philosophical traditions and power structures, marginalizing non-Western viewpoints (Birhane, 2024; Mohamed et al., 2020). Practical implications of decolonization include dealing with technology transfer agreements that sustain dependency, standards that disadvantage African countries, and developing indigenous technological capabilities, thereby reducing African countries' position as perpetual consumers of Western-developed applications, tools, and technologies.

Reducing technological dependency requires a shift from Africa primarily consuming external technologies to actively shaping, producing, governing, and benefiting from them. This, in turn, requires strengthening local innovation ecosystems, reforming global technology partnerships, expanding participation in global governance, and embedding African perspectives in technological ethics and design.

9.2.3 Algorithmic Bias and Fairness

Algorithmic bias—systematic and unfair discrimination embedded in automated decision systems—poses significant risks in Africa. Bias can originate from various sources: unrepresentative training data, algorithm design reflecting developers' unconscious assumptions, deployment in contexts different from training environments, and measurement systems that encode biased proxies.

Most AI and machine-learning systems are trained using datasets created in North America, Europe, or parts of Asia. These datasets mirror the social, economic, linguistic, and infrastructural conditions of those regions. When such systems are applied in African contexts—where languages, cultural practices, infrastructure conditions, and demographic patterns differ significantly—the algorithms may misinterpret data, generate inaccurate predictions, or reinforce systemic biases.

Machine learning systems mainly trained on data from developed countries have already demonstrated poor performance with African populations. Facial recognition technology has significantly lower accuracy for darker skin tones—error rates can reach up to 34% for dark-skinned women compared to 0.8% for light-skinned men (Buolamwini & Gebru, 2018). Speech recognition systems trained on American or British English struggle to recognize African accents and the code-switching common in multilingual African environments (Graham, 2024). Credit-scoring algorithms based on formal-sector data often unfairly disadvantage informal workers, who comprise 85% of employment in Africa (ILO, 2024).

Measurement systems often encode biased proxies when the variables used to quantify performance, risk, or opportunity indirectly reflect historical inequalities rather than the phenomenon they are intended to measure. Because many social, economic, and technological systems rely on observable indicators as substitutes for complex realities, designers select measurable variables—such as income, location, educational attainment, or digital behavior—as proxies for qualities like capability, creditworthiness, productivity, or trustworthiness. When these proxies are correlated with structural inequalities (e.g., race, geography, gender, or language), the measurement system inadvertently embeds those inequalities into its outputs.

As a result, the system appears objective while reproducing historical bias.

Table 9.1. Algorithmic Bias Risks, Affected Groups, and Recommended Governance Safeguards in Africa

Application Domain	Key Bias Risks	Affected Groups	Recommended Safeguards
Facial Recognition	34% error rate for dark-skinned women vs. 0.8% for light-skinned men	Women, darker-skinned populations	Mandatory accuracy testing across demographics; deployment moratorium until frameworks established
Credit Scoring	Formal-sector data bias; proxy discrimination	Informal workers (85% of employment), women, rural populations	Ban proxy discrimination; bias impact assessments; diverse training data
Hiring Algorithms	Historical bias in training data	Women, ethnic minorities, youth without formal credentials	Fairness audits; human oversight in final decisions; diverse training datasets
Medical Diagnosis	Training on European population data	African patients with region-specific conditions	Mandate African clinical trial data inclusion; locally validated models
Content Moderation	Excessive censorship of African languages	African-language speakers	Language-specific model testing; human review for contextual understanding
Agricultural AI	Models trained on non-African conditions	Smallholder farmers, rural communities	Locally validated models; participatory design with farmers

Source: Buolamwini & Gebru, 2018; African Digital Rights Network, 2024; UNESCO, 2024.

Algorithm design can also reflect developers' unconscious assumptions, values, and mental models, which influence how problems are framed, what data are used, and how outcomes are evaluated. Algorithms do not arise neutrally from mathematics; they are built through a series of design decisions —such as defining objectives, selecting variables, labeling training data, choosing optimization metrics, and setting thresholds. These decisions, often made in the United States, China, and Europe, inevitably mirror the perspectives and experiences of the people who create them. When development teams share similar educational, cultural, or socioeconomic backgrounds, certain assumptions about "normal" behavior, desirable outcomes, or relevant data may be embedded in the system without explicit awareness of differing contexts, especially in Africa.

Algorithmic fairness is a social and political issue that involves value judgments about acceptable trade-offs—since different fairness definitions can conflict—and Africa must develop contextually appropriate fairness standards rather than importing definitions from other regions (Research ICT Africa, 2024). Governance strategies for algorithmic bias include mandatory bias testing before deploying high-stakes systems, human oversight requirements, transparency obligations that enable external audits, and effective redress mechanisms. The African Union (2024) recommends that Member States establish requirements for algorithmic impact assessments of systems that affect fundamental rights.

Box 9.1. Comparative Insight: Algorithmic Governance

Canada's Algorithmic Impact Assessment (AIA) tool was introduced in 2019 as a mandatory requirement for federal government departments deploying automated decision systems. The AIA categorizes systems into four risk levels based on decision reversibility, the scope of legal impact, and the likelihood of discriminatory outcomes. Each level enforces progressively stricter requirements—including human-in-the-loop mandates, public notice, and external audits for the highest-risk categories. Canada's Directive on Automated Decision-Making also mandates that the source code of automated systems be provided upon request, establishing a transparency standard that African governance frameworks could adapt to enhance algorithmic accountability.

South Korea's AI Ethics Standards, adopted in 2020, articulate principles of human dignity, the public good, and technical safety and have been accompanied by mandatory algorithmic impact assessments for public-sector AI systems—a model that closely parallels Canada's approach. Japan's Social Principles of Human-Centric AI emphasize that AI must respect cultural diversity and promote inclusive societies, resonating with African concerns about culturally appropriate technology governance.

Colombia's AI Ethics Framework, developed through extensive public consultation including Indigenous and Afro-Colombian community participation, demonstrates how Latin American countries are incorporating diverse cultural perspectives into technology governance—an approach that parallels Africa's emphasis on Ubuntu-informed AI governance.

9.3 Data Protection, Privacy, and Cybersecurity

Data protection and cybersecurity are crucial for Africa's digital development. As societies become more digital, safeguarding personal information and defending systems from threats are vital to building trust, protecting rights, and expanding digital economies. The continent has made significant progress in creating data protection laws and developing cybersecurity expertise over the past decade; however, major challenges remain in implementation, enforcement, and institutional capacity building.

9.3.1 Data Protection Legal Frameworks Across Africa

Africa's data protection legal landscape has evolved considerably. In 2010, fewer than 10 African nations had comprehensive data protection laws. By 2025, 37 countries across Africa had enacted personal data protection legislation, including Algeria, Angola, Benin, Botswana, Burkina Faso, Cape Verde, Chad, Comoros, Republic of Congo, Côte d'Ivoire, Egypt, Eswatini, Ethiopia, Gabon, Ghana, Guinea, Equatorial Guinea, Kenya, Lesotho, Madagascar, Mali, Mauritius, Mauritania, Morocco, Mozambique, Niger, Nigeria, São Tomé and Príncipe, Senegal, South Africa, Tanzania, Togo, Tunisia, Seychelles, Uganda, Zambia, and Zimbabwe. This rapid legislative progress reflects a growing awareness of the importance of data protection, driven by the African Union's Data Policy Framework, demands from trading partners (notably the European Union), and persistent civil society efforts.

The African Union Convention on Cyber Security and Personal Data Protection (Malabo Convention, 2014) establishes continental standards that largely align with international frameworks, including the EU's General Data Protection Regulation (GDPR). However, only 16 of 55 AU member states have ratified the Convention by 2025.

The low ratification rate of the Malabo Convention highlights the limited resources for data protection activities across Africa. South Africa's Information Regulator—one of Africa's most well-equipped data protection authorities, with about 60 staff and a $5 million annual budget—still handles thousands of complaints and has limited capacity for proactive audits (South African Information Regulator, 2024). Most other African data protection authorities operate with even fewer resources.

The resource and enforcement capacity limitation also points to the fact that ambitious laws on paper often provide limited protection in practice. Research by the African Digital Rights Network (2024) found that 67% of small and medium-sized enterprises in Africa are unaware of their countries' data protection laws; among those aware, 52% do not understand the specific requirements, and 78% cite compliance costs as prohibitive. Public sector compliance is especially problematic. UNESCO (2024) surveys indicate that only 23% of Africans can name any data protection rights—a figure below 10% in rural areas.

Box 9.2. Comparative Insight: Data Protection in Canada, Latin America, and Oceania

Canada's data protection regime, anchored by the Personal Information Protection and Electronic Documents Act (PIPEDA) and the new Consumer Privacy Protection Act (CPPA), provides a model of federated privacy governance where federal and provincial laws coexist—relevant for Africa, where continental (Malabo Convention, AU Data Policy Framework), regional (REC-level), and national frameworks must function in concert. Canada's Office of the Privacy Commissioner, with a budget of approximately CAD 30 million and over 200 staff members, illustrates the scale of institutional investment required for effective data protection enforcement.

New Zealand's Privacy Act 2020, which grants the Privacy Commissioner the power to issue compliance notices and establishes mandatory breach notification, is considered one of the more modern privacy frameworks in the OECD. New Zealand's emphasis on cross-border data flow provisions—requiring that personal information sent overseas receive comparable protections—directly addresses data sovereignty concerns relevant to African debates.

Brazil's Lei Geral de Proteção de Dados (LGPD), modeled on the GDPR and enforced by the independent Autoridade Nacional de Proteção de Dados (ANPD), has quickly become the leading data protection framework in the Global South. The ANPD's step-by-step approach to building enforcement capacity—starting with guidance and education before moving to sanctions—provides a model for resource-limited African data protection environments.

9.3.2 Cybersecurity Threats and Capabilities

Cybersecurity threats impacting Africa have grown significantly in recent years. The African Union (2024) estimates that cybercrime costs African economies $4.2 billion annually—double the $2.1 billion reported in 2020. Major threats include financial fraud targeting mobile money and banking systems, ransomware attacks increasingly directed at critical infrastructure and healthcare, data breaches, business email compromise, and distributed denial-of-service attacks. By 2024, 38 African countries had operational national Computer Emergency Response Teams (CERTs), up from 18 in 2018, but most have only 3–8 technical staff and budgets under $300,000—insufficient for 24/7 monitoring, threat intelligence, incident response, and capacity building (African Union, 2024).

North Africa and Southern Africa lead in cybersecurity maturity, driven by relatively strong legal frameworks, solid technical infrastructure, and South Africa's well-developed cybersecurity ecosystem. West Africa, Eastern Africa, and Central Africa lag significantly behind. All regions score lowest in capacity building, highlighting the skills shortage discussed earlier. The African Development Bank (2024) estimates that only 15% of critical infrastructure operators have adopted strong cybersecurity controls aligned with international standards.

Box 9.3. Comparative Insight: Cybersecurity Governance in Asia, Latin America, and Oceania

Australia's Cyber Security Strategy 2023–2030 is supported by an investment of AUD 2.3 billion. Australia's Security of Critical Infrastructure Act imposes positive security obligations on operators and allows government assistance during major cyber incidents. New Zealand's National Cyber Security Centre and CERT NZ have developed cybersecurity capability through international partnerships. The Five Eyes intelligence alliance's cybersecurity cooperation among Australia, Canada, New Zealand, the UK, and the US highlights the importance of trust-based partnerships, a model that could guide REC's regional cybersecurity efforts.

Singapore's Cyber Security Agency and its Cybersecurity Act require compliance with the Cybersecurity Act for critical information infrastructure across 11 designated sectors. Singapore's approach to licensing cybersecurity service providers—mandating registration for penetration testers and managed security providers—helps ensure quality in a sector where Africa faces significant skills shortages.

The Organization of American States (OAS) Cybersecurity Programme has supported the creation and capacity-building of CERTs across Latin America and the Caribbean, offering a regional model for cooperation for Africa.

9.3.3 Privacy Rights and Surveillance Governance

Surveillance technologies highlight important conflicts between security priorities and privacy rights. Many African governments have implemented advanced surveillance tools, including communications interception, Internet monitoring, facial recognition, and social media surveillance. While there are valid security reasons for these measures, human rights groups report widespread monitoring targeting journalists, activists, opposition leaders, and civil society (Access Now, 2024; Privacy International, 2024). The African Commission on Human and Peoples' Rights (2024) found that fewer than 10 African countries have comprehensive surveillance laws that align with international human rights standards.

Internet shutdowns are the most extreme form of government digital control—and their frequency is increasing in Africa. Access Now (2024) reports 75 Internet shutdown incidents across 18 African countries in 2023, representing the largest share globally (40% of all shutdowns), despite Africa comprising only 17% of the world's population. Shutdowns cause substantial economic damage (estimated at $1.9 billion in 2023), disrupt healthcare, banking, and education systems, and violate international human rights laws.

9.4 AI Governance and Algorithmic Accountability

Among all emerging technologies, AI presents a significant ethical challenge for Africa. Artificial intelligence introduces specific governance issues: technical complexity, rapid development, opacity of decision-making, potential for widespread societal impacts, and the absence of established regulatory standards. African countries must create AI governance frameworks that support the beneficial use and growth of AI while minimizing risks such as discrimination, lack of transparency, and accountability gaps.

9.4.1 AI Ethics Principles for Africa

African AI governance frameworks are evolving around key ethical principles, shaped by international discussions and interpreted through African philosophical traditions. UNESCO's (2024) Recommendation on the Ethics of Artificial Intelligence, adopted by all African Union Member States, highlights fundamental principles that African countries are now incorporating into their local frameworks. Six core principles define AI ethics: fairness and non-discrimination; transparency and explainability; accountability; privacy and data protection; safety and security; and human agency and oversight—aligned with Ubuntu philosophy's emphasis on human relationships and collective decision-making.

Beyond these universal principles, African AI governance frameworks should emphasize four key areas.

- First, development-focused AI: unlike wealthy nations that can afford cautious regulation, African countries need AI to address urgent development challenges—such as agricultural productivity, healthcare access, educational quality, and climate adaptation.

- Second, data sovereignty: reliance on foreign AI systems creates strategic vulnerabilities.

- Third, linguistic and cultural diversity: Kenya's AI strategy emphasizes Swahili-language AI; Ghana emphasizes Akan, Ewe, and Ga; the AU Continental AI Strategy promotes pan-African multilingual research.

- Fourth, community-centered governance: approaches informed by Ubuntu promote community data trusts, participatory AI impact assessments, and community representation on oversight boards.

9.4.2 Algorithmic Accountability in Africa

Algorithmic impact assessments—systematic evaluations of AI systems' potential effects before deployment—are a crucial accountability tool. Canada's Algorithmic Impact Assessment framework (Box 9.1), which assesses systems across four risk levels based on decision reversibility, legal consequences, and fairness concerns, is being adopted by African countries. For example, Kenya's Data Protection Commissioner has proposed similar requirements for AI systems that affect rights and freedoms (Kenya Data Protection Commissioner, 2024). However, many developers lack the expertise to conduct meaningful assessments, methods for addressing Africa-specific harms are underdeveloped, and assessments often become mere formalities.

Algorithmic audits—independent evaluations of deployed systems—complement impact assessments by analyzing actual real-world performance. Audit costs (ranging from $50,000 to $500,000, depending on system complexity) and limited African auditing expertise present challenges.

Transparency requirements call for disclosures about AI systems—including registries of government AI deployments, documentation of training data and algorithms, disclosures when individuals interact with AI, and explanations of algorithmic decisions that affect individuals. Tunisia's Digital Economy Law requires government agencies to disclose their use of AI systems; South Africa's proposed AI regulation mandates the right to an explanation for major automated decisions; Morocco requires financial institutions to disclose key credit-scoring factors.

An effective redress mechanism is another ethical dimension that remains underdeveloped in Africa. The World Bank (2024) estimates that fewer than 2% of Africans experiencing algorithmic harms ever seek formal redress—due to unawareness of their exposure to algorithmic decisions, limited knowledge of appeal rights, burdensome processes, long backlogs at oversight bodies, and limited remedies even when violations are identified.

9.4.3 National AI Strategies and Governance Frameworks

National AI strategies offer comprehensive approaches towards promoting AI benefits, while also managing its risks. By 2025, fourteen African countries—Benin, Côte d'Ivoire, Democratic Republic of the Congo, Egypt, Ethiopia, Ghana, Lesotho, Kenya, Mauritius, Rwanda, Senegal, South Africa, Zambia, and Zimbabwe—have developed dedicated national AI strategies or policies. Common objectives include building national data infrastructure, strengthening digital ecosystems, encouraging AI-driven innovation in key sectors such as agriculture, healthcare, education, manufacturing, mining, and public administration, and establishing regulatory frameworks to address risks related to bias, privacy, and accountability.

Institutional mechanisms such as national AI councils, AI institutes, and Centers of the Fourth Industrial Revolution have been established to oversee implementation and uphold ethical standards. Along with these fourteen countries, Algeria, Mauritania, Morocco, Nigeria, Tunisia, Tanzania, and Uganda are promoting AI through research strategies, institutional reforms, and digital transformation initiatives. However, in most countries, these strategies lack the necessary skills, financial and technical resources to fully support their implementation.

Box 9.3. Comparative Insight: AI Governance in Canada, China and India

Canada's Artificial Intelligence and Data Act (AIDA), introduced as part of the Digital Charter Implementation Act, is among the first efforts worldwide to establish binding AI legislation. AIDA creates a risk-based framework for high-impact AI systems, mandates responsible AI practices, including bias mitigation and transparency, and establishes the Office of the AI and Data Commissioner. Canada's Pan-Canadian Artificial Intelligence Strategy, supported by over CAD 2 billion, has established three national AI institutes—Mila in Montreal, the Vector Institute in Toronto, and Amii in Edmonton—serving as a model for how public investment can boost AI research ecosystems that African countries could replicate at national and regional levels.

China's approach to AI governance—which combines ambitious development targets with sector-specific regulations covering algorithmic recommendations, deep synthesis (deepfakes), and generative AI—demonstrates how a major developing economy can pursue rapid AI advancement while establishing regulatory guardrails.

India's National AI Strategy (NITI Aayog) focuses AI deployment on five priority sectors—healthcare, agriculture, education, smart cities, and transportation—mirroring African development needs and demonstrating how AI governance can be structured to address development challenges rather than simply managing risks in isolation- a model adopted by the AU Continental AI Strategy.

9.5 Standards, Regulation, and Regional Harmonization

Technical standards and regulatory approaches fundamentally shape how emerging technologies are developed and used. Standards ensure compatibility, set quality benchmarks, and promote trade. Regulatory frameworks specify permitted activities, assign responsibilities, and protect public interests.

9.5.1 Technology Standards and African Participation

Global technology standards are primarily shaped by international organizations—ISO, IEC, ITU, and IETF—and African involvement in these organizations remains notably limited. African countries hold about 4% of leadership roles in standards-setting bodies, even though they represent 17% of the global population and 28% of UN members (ITU, 2024). African technical experts participate in roughly 12% of working groups that develop telecommunications and information technology standards; African companies and research institutions contribute to fewer than 8% of technical proposals.

Consequences of this limited participation include the availability of standards that do not adequately serve African needs: mobile telecommunications standards that favor high-capacity urban networks over affordable rural coverage; cybersecurity standards that assume resources are available and that do not conform to many African settings; and AI fairness standards that may not address biases affecting African populations.

The African Organization for Standardization (ARSO), with an annual budget of about $2 million (compared to ISO's over $100 million), manages regional standards development. The African

Continental Free Trade Area's Digital Trade Protocol aims to establish mechanisms to harmonize standards across continental markets (UNCTAD, 2024). This calls for substantial resources to establish context-specific standards, participate in global standard setting and enforce them across the continent.

9.5.2 Regulatory Approaches for Emerging Technologies

Emerging technologies require stable regulatory environments. As discussed throughout this volume, regulatory sandboxes—allowing companies to test innovative products or services under regulatory oversight with relaxed rules—have become increasingly common across Africa, inspired by the UK Financial Conduct Authority's 2016 pioneering model. Kenya's Capital Markets Authority, along with those of South Africa, Mauritius, Rwanda, and Nigeria, operates a fintech and digital innovation sandbox.

Figure 9.2. Comparison of Regulatory Approaches for Emerging Technologies in Africa.

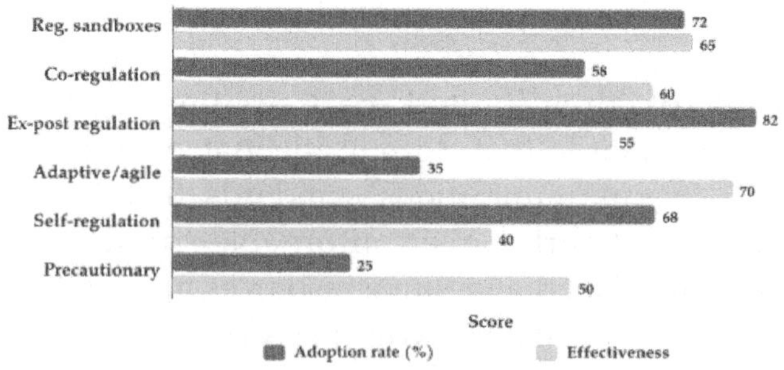

Source: WEF (2024); World Bank (2024)

Co-regulation and regulatory sandboxes offer the best balance of effectiveness and manageability in regulating new technologies(World Bank, 2024). Ex-post regulation has the highest adoption rate in Africa, but moderate effectiveness.

Adaptive and agile regulation—recognizing that frameworks must change as technologies and their effects become clearer—offer promising options.

Box 9.4 Comparative Insight: Regulatory Innovation in Canada and Singapore

Canada's experience with regulatory sandboxes extends beyond fintech to encompass energy (the Canada Energy Regulator's Innovation Sandbox), transportation (autonomous vehicle testing frameworks in Ontario and Quebec), and healthcare (Health Canada's regulatory pathway for AI-enabled medical devices)—demonstrating how the sandbox model can be adapted across sectors.

Singapore's regulatory sandbox model—run by the Monetary Authority of Singapore (MAS)—is widely seen as one of the most effective globally and has directly influenced the design of sandboxes in Kenya, Rwanda, and South Africa. Singapore's sandbox-express pathway, which allows low-risk innovations to be tested within 21 days of application, demonstrates how regulatory agility can align with startup timelines.

9.5.3 Regional Harmonization

Africa has initiated regional efforts to unify policies and regulations for managing the cross-border benefits and risks of emerging technologies. Digital payments exhibit a high degree of harmonization, particularly among the most coordinated countries, driven primarily by the need for mobile money integration and regional payment systems. Cybersecurity harmonization is shaped by mutual recognition and regional CERT networks. AI ethics and content moderation are still in the early stages of policy development.

Regional Economic Communities serve as important tools: the East African Community has unified telecommunications rules and data protection laws; SADC has developed a Model Law on Data Protection; ECOWAS operates a regional CERT and has issued guidelines on electronic signatures and consumer protection.

Figure 9.3. Regional Policy Harmonization Progress by Domain in Africa

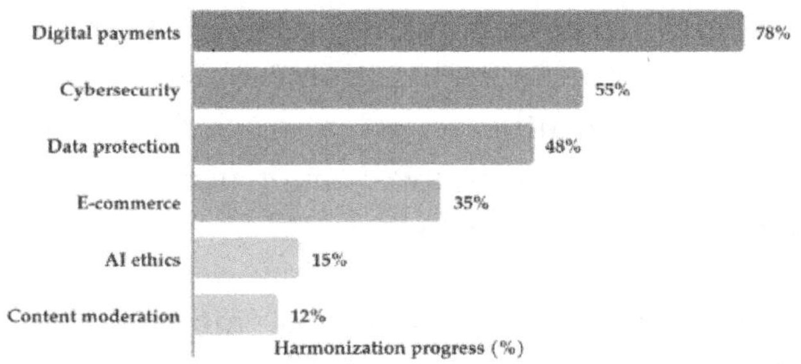

Source: African Union (2024); UNCTAD (2024); EAC (2024); ECOWAS (2024).

However, multiple challenges hinder efforts to harmonize at both continental and regional levels: different legal systems (civil versus common law), language barriers, varying regulatory capabilities, sovereignty concerns, and conflicting national priorities. To overcome these obstacles, it is important to start with less contentious areas—such as technical standards, payment interoperability, and cybersecurity cooperation—to build momentum; adopt flexible harmonization approaches that accommodate national differences within shared frameworks; and utilize concrete pilot projects to garner political support (UNCTAD, 2024).

9.6 Digital Rights, Internet Freedom, and Integrated Governance Strategies

Digital rights—the extension of human rights into digital spaces—serve as the moral foundation for technology governance. Rights to privacy, freedom of expression, freedom of assembly, and access to information must be protected as societies adopt new technologies. However, African governments are increasingly restricting digital rights through surveillance, content regulation, and Internet shutdowns. The African Declaration on Internet Rights and Freedoms (2014), although not legally binding, provides an influential normative framework affirming that existing human rights also extend to digital spaces.

From the rights perspective, effective technology governance requires collaboration among diverse stakeholders—including governments, the private sector, civil society, the technical community, academia, and users. No single actor has enough knowledge, authority, legitimacy, or resources to govern technology alone. Several principles should guide African technology governance: proportionality (interventions corresponding to risk levels); transparency (openness about governance processes and decisions); accountability (clear responsibilities and consequences for failures); inclusion (incorporating the perspectives of women, rural populations, linguistic minorities, youth, and persons with disabilities); and sustainability (frameworks that adapt to technological changes while providing stability for planning and investment).

Box 9.5 Comparative Insight: Digital Rights in Asia, Latin America, Canada, and Oceania

Canada's Charter of Rights and Freedoms, as interpreted by courts in the digital context, provides constitutional protection for privacy and expression online. Canada's Office of the Privacy Commissioner has issued guidance on facial recognition, AI decision-making, and children's privacy that has influenced other jurisdictions. Canada's experience with the Online Harms Act, which seeks to address online safety while preserving freedom of expression, illustrates the complex balancing act that African countries face as they develop content regulation frameworks.

Australia's Online Safety Act 2021, which established the eSafety Commissioner as a dedicated regulator for online safety, provides a practical model for addressing online harms through a specialized agency.

Japan's Act on the Protection of Personal Information, combined with its strong constitutional protections for privacy and expression, illustrates how Asian democracies integrate digital rights into existing legal frameworks rather than creating parallel systems. Brazil's Marco Civil da Internet (Internet Bill of Rights, 2014) is widely regarded as one of the most progressive internet governance frameworks globally, establishing principles of net neutrality, freedom of expression, and privacy online that have influenced legislation across the developing world.

Overall, the analysis of emerging technology risks and governance indicates that Africa has made significant progress in key areas, but challenges remain. Data protection leads the legislative efforts, with 37 countries now having comprehensive laws—up from fewer than 10 in 2010. Cybersecurity capacity is expanding, with 38 countries operating national CERTs, and AI governance is gaining ground, as 14 countries have published dedicated strategies aligned with the AU Continental AI Strategy.

Digital payments harmonization is the most advanced regional effort, driven by the need for mobile money integration, and the African Declaration on Internet Rights and Freedoms offers a well-established normative framework for digital rights. These achievements showcase a continent actively developing the governance structures needed for responsible technology adoption.

However, significant gaps remain between policy ambition and practical enforcement across all governance areas. Data protection authorities are severely underfunded. Cybercrime costs the continent $4.2 billion each year, yet most national CERTs only have 3–8 staff members and budgets under $300,000.

AI strategies and governance face implementation gaps, with limited technical regulatory capacity and fewer than 2% of those impacted by algorithmic harms seeking formal redress.

African participation in global standards bodies remains minimal, at 4% of leadership roles; AI ethics harmonization is still in the early stages; and 75 Internet shutdowns in 2023—40% of the global total—highlight an ongoing digital rights challenge. Bridging these gaps requires ongoing investment in institutional capacity, faster implementation of the AfCFTA Digital Trade Protocol, and a move from legislative adoption to effective enforcement and increased public awareness.

Table 9.2. Summary of Emerging Technology Governance in Africa

Governance Area	Status	Key Achievement	Main Gap	Recommended Action
Data Protection	37 countries with laws	Rapid legislative expansion	Enforcement capacity; SME awareness (67% unaware)	Increase institutional investment; public awareness campaigns
Cybersecurity	38 countries with CERTs	Growing institutional capacity	$4.2B annual losses; staff/budget shortages	Priority national security investment; regional cooperation
AI Governance	14 countries with strategies	Continental AI Strategy adopted	Implementation gap; limited technical regulatory capacity	Dedicated AI institutes; pilot programs; inter-agency coordination
Algorithmic Accountability	Emerging frameworks	Impact assessment adoption beginning	<2% pursue redress; limited auditing capacity	Mandatory assessments for high-risk systems; accredited auditors
Standards	ARSO operational	African agricultural and mobile money standards	4% of global leadership roles; 8% of technical proposals	Strategic participation in ISO/IEC/ITU; capacity building
Regional Harmonization	Variable by domain	Digital payments highly harmonized	AI ethics and content moderation nascent	Accelerate AfCFTA Digital Trade Protocol
Digital Rights	Declaration adopted	Normative framework established	75 shutdowns in 2023; limited surveillance laws	Shutdown moratoria; judicial oversight; digital rights legislation

Source: African Union, 2024; UNESCO, 2024; UNCTAD, 2024.

9.7 Conclusion

Governing emerging technologies is a top development priority for Africa. Technology can accelerate progress toward development goals, improve government efficiency, create more economic opportunities, and enhance quality of life. However, poorly managed technology can increase inequality, violate rights, create strategic dependencies, and enable new forms of exploitation and control. The governance choices African countries make now will shape technological paths for many years. Several key insights emerge:

- First, Africa is rapidly strengthening its technology governance capacity, and regional harmonization is progressing.

- Second, a persistent gap exists between policy goals and enforcement, driven by capacity limits, resource shortages, and political obstacles.

- Third, African perspectives—such as Ubuntu philosophy, community-based ethics, decolonial thought, and development goals—must genuinely shape governance frameworks, rather than African nations simply endorsing norms imposed from outside.

- Fourth, involving multiple stakeholders is essential.

- Fifth, algorithmic accountability, data sovereignty, and digital rights are vital to ensuring technology supports African development rather than increasing dependence.

Experiences from Asia, Latin America, the Caribbean, Canada, Australia, and New Zealand offer valuable additional insights: Canada's Algorithmic Impact Assessment and federated privacy governance model; Australia's comprehensive cybersecurity strategy and online safety framework; Singapore's AI Verify testing framework and fintech sandbox model; India's data protection legislation and AI strategy focused on development priorities; Brazil's progressive internet governance (Marco Civil) and AI legislation and Colombia's inclusive AI ethics framework—all provide practical tools that African governance frameworks can adapt. Importantly, these diverse experiences demonstrate that effective technology governance is an ongoing process of iteration and learning, and that countries at all income levels are grappling with similar challenges—a principle directly relevant for African countries building governance capacity in rapidly evolving technological landscapes.

Finally, priorities for strengthening African technology governance include: significantly investing in regulatory capacity; encouraging regional cooperation and harmonization through the AfCFTA Digital Trade Protocol and RECs; ensuring African participation in global standards and governance forums; protecting digital rights by preventing surveillance overreach and ending Internet shutdowns; establishing multi-stakeholder governance institutions with clear mandates and resources; and raising public awareness of technology governance as a development priority.

References

Access Now. (2024). The KeepItOn report: Internet shutdowns in 2023. Access Now Publications.

African Commission on Human and Peoples' Rights. (2024). Resolution on freedom of expression and access to information. ACHPR Resolutions.

African Development Bank. (2024). Cybersecurity in Africa: State of play. AfDB Digital Development Department.

African Digital Rights Network. (2024). Data protection compliance study. ADRN Research Papers.

African Organisation for Standardisation. (2024). African standards development report. ARSO Publications.

African Union. (2024). Continental's Artificial Intelligence Strategy. African Union Commission.

African Union. (2024). Convention on cyber security and personal data protection: Implementation status. AU Legal Department.

Birhane, A. (2024). Algorithmic colonisation of Africa. In P. Taylor (Ed.), Decolonising AI. Oxford University Press.

Buolamwini, J., & Gebru, T. (2018). Gender shades: Intersectional accuracy disparities. Proceedings of Machine Learning Research, 81, 1–15.

East African Community. (2024). EAC digital integration framework: Progress report. EAC Secretariat.

ECOWAS Commission. (2024). Regional cybersecurity and data protection harmonisation. ECOWAS Publications.

Graham, M. (2024). Digital work in Africa: Algorithmic imperialism. International Labour Review, 163(1), 45–62.

International Labour Organization. (2024). Informal employment and digital platforms in Africa. ILO Research Department.

International Telecommunication Union. (2024). Global Cybersecurity Index 2024. ITU Publications.

Kaspersky. (2024). Cybersecurity threat landscape: Africa. Kaspersky Security Bulletin.

Kenya Data Protection Commissioner. (2024). Guidelines on algorithmic systems. Office of the Data Protection Commissioner.
Mohamed, S., Png, M. T., & Isaac, W. (2020). Decolonial AI. Philosophy & Technology, 33(4), 659–684.
Personal Data Protection Commission Singapore. (2024). Model AI governance framework. PDPC Publications.
Privacy International. (2024). State surveillance in Africa: Technologies and practices. PI Research.
Research ICT Africa. (2024). AI governance in Africa: Policy and practice. RIA Policy Paper.
South African Department of Communications. (2024). National artificial intelligence strategy. Government of South Africa.
South African Information Regulator. (2024). Annual report 2023/2024. Information Regulator South Africa.
UNCTAD. (2024). Digital regulation in Africa: Survey of practices. UNCTAD Technical Papers.
UNECA. (2024). Technology sovereignty for Africa. UNECA Policy Brief.
UNESCO. (2024). AI ethics and fairness in Africa. UNESCO Digital Society Division.
UNESCO. (2024). Recommendation on the ethics of artificial intelligence. UNESCO Publishing.
World Bank. (2024). Regulatory sandboxes: Global practices and lessons for Africa. World Bank FinTech Note.
World Economic Forum. (2024). Global risks report 2024: Technology risks. WEF Publishing

PART V

Pathways for Africa's Transformation through Accelerated Adoption of Emerging Technologies

Chapter 10: Strategic Pathways for Africa's Emerging Technology Transformation

10.1. Introduction

The previous chapters have established that Africa's technological future depends not just on which technologies are chosen but also on the strength of the foundations and the quality of governance that support them. Infrastructure quality accounts for 40–60% of the variation in technology adoption rates, while institutional quality accounts for an additional 35–50%.

Financing remains a key barrier to the adoption of emerging technologies for development. The continent faces an annual infrastructure financing shortfall of $68–108 billion. A governance preparedness crisis (average score of 36 compared to a risk severity of 84), skills shortages across all technology sectors, and market fragmentation across many African countries, with hundreds of distinct regulatory requirements, also hinder progress. Yet, Africa also shows impressive innovation—mobile money, drone delivery, pay-as-you-go solar—and possesses genuine comparative advantages in critical minerals, renewable energy, linguistic diversity, and demographic growth.

This chapter translates the book's analysis into actionable strategies, providing implementation roadmaps at continental, regional, and—most importantly—national levels. While continental strategies via the African Union offer broad frameworks and regional integration through RECs creates scale, national governments hold primary responsibility for creating the conditions that determine whether emerging technologies deliver inclusive development or exacerbate existing divides. Accordingly, this chapter focuses mainly on what countries must do: design and implement comprehensive national digital transformation strategies, prioritize and sequence investments, build human capital, mobilize financing, strengthen governance, foster innovation ecosystems, and incorporate inclusion as a core principle across all technology initiatives.

10.2 Continental and Regional Coordination

The African Union's Digital Transformation Strategy for Africa (2020–2030) sets ambitious goals: universal, affordable broadband, digital identification for all Africans, digital payment accounts for 80% of adults, one million digital jobs, and a continental digital single market—with an estimated $600 billion in financial flow over the decade. Five years into the plan, progress is mixed: mobile broadband coverage has increased from 60% to 80%, digital payment adoption has grown from 40% to 55% of adults, but skills development lags behind all targets, digital government rollout is inconsistent, and actual investment reached only $38 billion in 2023—63% of the $60 billion annual goal.

The AfCFTA Digital Trade Protocol, adopted in February 2024, is the most critical continental instrument for reducing market fragmentation. The Pan-African Payment and Settlement System (PAPSS) has processed over $180 million in cross-border payments since its 2022 launch (African Export-Import Bank, 2024).

Regional Economic Communities serve as the main mechanisms for translating continental goals into practical integration. The East African Community's removal of roaming charges in 2023 and its East African Payment System ($8.5 billion in transactions that year) show what phased, consensus-based integration can accomplish. ECOWAS's West Africa Broadband Network has completed 60% of the planned routes, with financial support from public, private, and development partners, including the World Bank. SADC's Integrated Regional Electronic Settlement System enables real-time gross settlement among member central banks. ASEAN's Digital Masterplan 2025 and the EU's Digital Single Market demonstrate that harmonizing this complexity is achievable within 5–8 years with consistent effort.

10.2.1 The Role of the AU and RECs in Promoting Continental Public Goods

Technology adoption reaches its full potential when supported by continental public goods—shared infrastructure, harmonized standards, and collaborative frameworks that no single country can efficiently provide alone. The African Union and Regional Economic Communities play indispensable roles in creating and sustaining these public goods across four critical domains.

Pan-African Digital Infrastructure

Continent-wide digital infrastructure is a crucial public good that underpins all technological applications. The AU and RECs play vital roles in expanding cross-border backbone networks; supporting Internet Exchange Points that keep intra-African traffic local instead of routing through Europe or North America; promoting pan-African cloud infrastructure and data centers hosting African content and applications within the continent; and advancing spectrum harmonization to enable economies of scale and device manufacturing. By coordinating investments and establishing common technical standards through projects like PIDA, the AU needs to continue playing a key role in ensuring that national infrastructure efforts link to a continent-wide digital backbone. At the same time, RECs such as the EAC, IGAD, ECOWAS, ECCAS, AMU, and SADC facilitate cross-border connectivity projects within their respective subregions.

Shared Research and Innovation Facilities

Africa's research infrastructure, including compute resources, remains significantly underdeveloped relative to the continent's needs and potential. The AU and RECs could play a catalytic role in establishing and sustaining shared facilities—such as continental and regional centers of excellence in AI, IoT, renewable energy, and other frontier technologies—that provide the critical mass of equipment, expertise, and collaboration that individual countries cannot provide on their own. The AU's role in brokering these investments and the RECs' role in promoting the hosting regional nodes are essential for making shared research infrastructure viable.

Continental Data Governance Frameworks

Data governance is arguably the most important public good on the continent in the digital age. The AU Data Policy Framework and its implementation plan aim to strengthen and harmonize data governance across Africa, creating a shared data space that balances openness with sovereignty. The AU provides the normative architecture, while RECs serve as implementation platforms for subregional harmonization.

Technology Standards Harmonization and Cross-Border Regulatory Cooperation

Fragmented technical standards and regulatory approaches across African countries pose a major barrier to deploying emerging technology at scale. Standards harmonization—for IoT devices, AI systems, blockchain applications, drone operations, health technologies, and digital financial services—is a public good that lowers costs, enables scaling, and accelerates adoption. RECs serve as natural platforms for gradual harmonization, with successful regional efforts expanding to the continental level under AU coordination. Harmonizing technology standards and fostering cross-border regulatory cooperation are therefore essential, and the AU's convening power, combined with the RECs' closer ties to national regulators, makes this layered approach the most practical path forward.

For national governments, the implication is clear: aligning national strategies with continental and regional frameworks—rather than pursuing isolated approaches—multiplies the impact of every domestic investment. The sections that follow focus on how countries can achieve this.

10.3 National Digital Transformation Strategies

10.3.1 Designing Comprehensive National Strategies

Transformation should happen at the national level. Every African country must develop or update a comprehensive national digital transformation strategy that addresses both the opportunities and risks of emerging technologies. The evidence from this book is clear: countries with coherent, well-resourced strategies—such as Rwanda, Kenya, Mauritius, Morocco, Egypt, and South Africa—consistently outperform those without them, regardless of income level.

Effective strategies arise from inclusive, evidence-based processes. Kenya's Digital Economy Blueprint involved over 1,200 stakeholders over 18 months; Rwanda's strategy was developed based on household surveys, business assessments, infrastructure mapping, and regulatory gap analysis; South Africa used multi-criteria decision analysis with transparent documentation of trade-offs.

Core components of successful national frameworks include a clear vision statement aligned with broader national development plans — such as Kenya's Digital Strategy, which aligns with Vision 2030, and Rwanda's Smart Rwanda Master Plan, which aligns with Vision 2050. Egypt's National AI Strategy (2025–2030) provides another example of a standalone strategy dedicated to emerging technologies. The Strategy combines strong institutional governance with a structured framework based on six pillars: governance, data and infrastructure, technology development, ecosystem growth, talent development, and sectoral applications. Through coordinated policy actions, pilot projects, talent development, and measurable targets, the National AI Strategy aims to build a competitive AI industry and integrate AI across government and economic sectors by 2030.

Most existing strategies in Africa lack sufficient provisions for emerging technologies—AI, blockchain, IoT, and robotics—necessitating updates to address both their development potential and governance challenges. Without a dedicated implementation authority with adequate staffing and funding, strategies risk remaining aspirational documents rather than operational programs.

10.3.2 Prioritizing, Sequencing, and Coordinating Across Government

With limited resources, African countries often face tough choices about technology priorities. The book's analysis provides clear guidance: prioritize based on development impact, implementation feasibility, cost-effectiveness, and strategic alignment—while sequencing investments to account for prerequisite dependencies. Digital government services need basic connectivity; AI applications require data infrastructure and governance; advanced manufacturing depends on reliable electricity. Solar home systems ($100–300 per household) offer immediate benefits at low cost. Extending the grid to sparsely populated areas ($5,000+ per household) is less cost-effective but may be justified on equity grounds. AI research might have a limited short-term impact, but it helps build long-term capabilities.

Zambia's Digital Transformation Strategy (2023-2027), which aligns with the 8th National Development Plan, is being implemented through three sequential phases—foundation, acceleration, and innovation—and provides a clear example of sequencing. The foundation phase emphasizes establishing the essential enablers of digital transformation, including expanding broadband infrastructure, developing digital government platforms, strengthening regulatory and cybersecurity frameworks, and introducing foundational systems such as digital identity and interoperable government infrastructure. The acceleration phase focuses on large-scale adoption and integration of digital technologies across government and key economic sectors by digitizing public services, expanding digital financial services, strengthening digital skills, and deploying sector-specific solutions in agriculture, health, education, and trade. The innovation phase aims to develop a mature digital economy by promoting technology entrepreneurship, expanding innovation ecosystems, supporting emerging technologies such as AI and data-driven services, and integrating Zambia into regional and global digital value chains (Government of Zambia, 2023).

Countries should balance quick wins—such as digital business registration, mobile money integration, agricultural SMS platforms, and telehealth services—with foundational investments in national fibre backbones, digital identity systems, data governance frameworks, and research institutions. South Africa explicitly allocates 30% of its technology budget to quick wins and 70% to foundational investments.

Whole-of-government coordination addresses the institutional fragmentation identified by the United Nations (2024) as a major barrier to effective digital government in Africa. Estonia's "once-only" principle—collecting information once and sharing it across agencies—has been adopted by Rwanda through an integrated government data architecture. Other coordination mechanisms include high-level inter-ministerial committees (Nigeria's NITDA Governing Board), dedicated digital transformation agencies (Tunisia's Digital Governance Agency), Chief Digital Officers (Mauritius), and shared service centers (Ghana's Government Cloud Infrastructure, which cut costs by roughly 40%).

India's comprehensive digital coordination through the Ministry of Electronics and IT, which oversees the India Stack ecosystem serving 1.4 billion people, shows how national institutional coherence can generate widespread benefits in identity, payments, and data exchange. Canada's Canadian Digital Service and Australia's Digital Transformation Agency serve as examples of how dedicated agencies can modernize government services through agile, user-centered design.

10.4 Building National Foundations

The key insight in the book is that foundations — not technologies — drive successful adoption. Technology adoption follows a clear hierarchy: physical infrastructure supports digital connectivity, which then enables digital services. All of this depends on human capital and is guided by governance.

Attempting advanced innovations without the necessary foundational elements, especially energy and connectivity, often results in poor outcomes. Only 47% of Africa's population has Internet access (compared to 67% worldwide); 600 million Africans lack electricity; and businesses face over 235 hours of outages each year.

Figure 10.1. Critical Success Factors: Importance vs. Strength

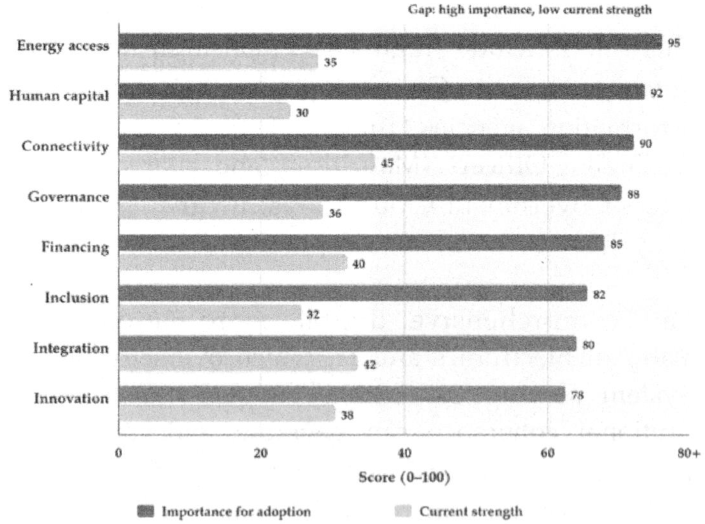

Source: AfDB (2024); WEF (2024).

10.4.1 National Energy Strategies for Technology

Energy access remains the biggest obstacle to the adoption of emerging technologies. National energy policies should prioritize both access and reliability by adopting a range of strategies. Morocco's development of 3,700 MW of renewable capacity—aiming for 52% renewable electricity by 2030—shows effective policy design, including measures such as regulatory certainty, government-backed off-take guarantees, and currency-hedging tools for foreign investors.

For rural electrification, countries should expand distributed solutions such as pay-as-you-go solar systems, mini-grids, and community-owned renewable microgrids. Canada's success in ending reliance on diesel in remote Indigenous communities through solar microgrids and Australia's ARENA-funded remote microgrid program serve as models for community-based electrification in Africa. Supportive policies, blended financing mechanisms, and stronger institutional capacity are crucial for ensuring reliability, sustainable maintenance, and productive use of electricity in rural communities.

10.4.2 National Connectivity Strategies

African governments must bridge the connectivity gap through a mix of regulatory reform, public investment, and private-sector mobilization. Mobile broadband covers 84% of sub-Saharan Africa's population, but only 28% actually use mobile Internet in 2024—a 56-percentage-point gap primarily caused by affordability issues (1 GB costs over 9% of monthly income in the least affordable markets) and device costs (an average of $120 for smartphones, or 23% of annual income in low-income countries).

Countries should promote infrastructure competition by ensuring transparent spectrum allocation and licensing, investing in national fiber backbones that connect submarine cable landing points to secondary cities and rural areas, developing and implementing broadband strategies that support multiple last-mile solutions such as fixed wireless access, community networks, and LEO satellite services in remote locations, and offering demand-side subsidies for low-income populations. Morocco's telecommunications liberalization attracted $12 billion in private investment. It achieved 95% mobile coverage by establishing clear rules, guaranteeing regulatory independence, and setting universal service obligations—serving as a model for the continent.

Data infrastructure requires parallel national investment. Only 2.5% of the global data centers are located in Africa. Furthermore, only around 50 Internet Exchange Points are active, causing much intra-African traffic to be routed internationally. Countries should prioritize establishing or expanding national IXPs and building data centres through public-private partnerships.

10.5 National Human Capital Development

Human capital deficits are among Africa's most serious challenges. Only 28% of Africans possess basic digital skills. The continent needs 9 million more software developers by 2030, but it produces just 200,000 technology graduates ready for work each year—resulting in an annual gap of 700,000. Cybersecurity has 15 open positions for every qualified professional, while AI and data science have 12. Brain drain worsens the skilled human resource pool: 35% of Africa's tech professionals work abroad, and talent is being drawn away by salary differences of 8–15 times.

10.5.1 National Digital Skills at Scale

Countries must pursue a three-track approach to skills development.

- Track one emphasizes foundational digital literacy for the general public by integrating computer literacy into primary and secondary curricula (e.g., Rwanda's Competency-Based Curriculum requires computer literacy from primary school; 89% of secondary schools now offer computer education); distributing devices with pedagogical support (Kenya's Digital Literacy Programme reached 1.2 million students); and making training materials accessible in local languages—a critical barrier since most interfaces are in English, French, or Arabic.

- Track two enhances intermediate technical skills through vocational training, bootcamps, and alternative pathways such as Andela's training-and-placement model, which has trained over 5,000 engineers with an 85% placement rate. Moringa School has achieved over 80% employment rates, and ALX/Holberton offers tuition-free engineering education. These models demonstrate scalable skill development options but must expand beyond software to include cybersecurity, data science, cloud computing, and AI, where shortages are most severe.

- Track three builds advanced research and innovation capacity through universities and research institutions. Africa invests only 0.4% of its GDP in research and development, compared to 1.8% globally and 2–4.5% in leading Asian economies. Countries should expand competitive research and development grant programs aligned with development priorities; strengthen university–industry partnerships through professional technology transfer offices, revenue-sharing incentives, and structured internships; and develop or reinforce centers of excellence in key technology areas.

South Korea's transformation through sustained education investment (2.6% to 5.1% of GDP over three decades), Australia's Cooperative Research Centres program (three decades of university–industry collaboration), and Canada's Scientific Research and Experimental Development tax incentive for private-sector research and development demonstrate how targeted human capital strategies—pursued with patience over decades—can yield transformative results.

10.5.2 Addressing Brain Drain

National strategies for brain drain must address both push and pull factors:

- On the pull side, African countries should focus on raising local salaries and expanding career opportunities for technology professionals, developing world-class research environments that attract and keep talented people, and ensuring that the regulatory and business climate supports the growth of technology careers domestically.

- On the proactive side, countries should engage the diaspora through structured programs for knowledge transfer, investment, and periodic return, creating circular migration pathways that allow professionals to gain international experience and return with improved skills and networks. They can also leverage diaspora remittances through diaspora bonds, venture capital funds, and angel investment networks.

Canada's Global Skills Strategy—which fast-tracks work permits for highly skilled workers while maintaining strong immigration pathways—demonstrates how countries can attract global talent and retain domestic talent.

10.6 National Technology Governance and Regulation

The book highlights a systemic governance preparedness crisis: across eight technology risk categories, the average preparedness score is only 36, while the average severity is 84, and the likelihood is 82. Thirty-seven countries now have data protection laws, 14 have published AI strategies, and 38 operate national CERTs—but implementation consistently lags legislation.

Governance quality accounts for 35–50% of the variation in technology adoption rates; countries with scores above 65 on governance quality have 2–3 times higher technology penetration than those with scores below 40. National governance strategies must build institutional capacity, establish appropriate regulatory frameworks, and protect digital rights.

10.6.1 Building National Regulatory Capacity

Regulatory capacity is vital for accelerating the adoption of emerging technologies. Data protection authorities are significantly underfunded in Africa: most have fewer than 10 employees and budgets of less than $500,000. Canada's Office of the Privacy Commissioner (more than 200 staff, CAD 30 million budget) demonstrates the scale required.

Countries must invest in recruiting and retaining technically skilled regulators, even if it means competing with private-sector salaries. They should also provide ongoing training in rapidly evolving technology fields, establish independence from political interference, and strengthen enforcement capabilities. Regulatory sandboxes have proven effective: Kenya, Rwanda, South Africa, Mauritius, and Nigeria all operate sandboxes. Singapore's sandbox-express pathway (21-day approval for low-risk innovations), Australia's Security and Investment

Commission's enhanced sandbox (24-month testing), and Brazil's Central Bank sandbox demonstrate flexible models.

10.6.2 National AI Governance

Fourteen African countries have published national AI strategies in 2025, but closing the gap between strategy and implementation requires dedicated institutions, development use cases, and technical regulatory capacity. Countries should establish national AI governance frameworks and institutions or centers of excellence to coordinate research, provide policy advice, build capacity, support the conduct of mandatory algorithmic impact assessments for high-risk AI systems that affect fundamental rights and establish government AI deployments registries and rules for the right to explanations for automated decisions. They should also develop accessible redress mechanisms against algorithmic harms and enable users to pursue formal remedies.

AI governance must be grounded in African ethical traditions—such as Ubuntu's emphasis on communal well-being, collective data ownership, and community accountability—while also integrating international frameworks. Brazil's AI legislation protecting vulnerable populations, New Zealand's Algorithm Charter, including Māori rights, and Singapore's AI Verify testing toolkit each provide adaptable elements.

10.6.3 National Cybersecurity Strategies

Cybercrime costs African economies $4.2 billion annually, yet most national CERTs operate with only 3–8 staff and budgets under $300,000. Countries need to establish or strengthen national CERTs with sufficient staffing, budgets, and mandates for 24/7 monitoring and incident response; require cybersecurity programs for critical infrastructure operators (only 15% have adopted strong controls); develop sector-specific cybersecurity requirements for telecommunications, finance, energy, education, and health; and invest in cybersecurity workforce development.

Australia's Cyber Security Strategy 2023–2030 (AUD 2.3 billion), Canada's Canadian Centre for Cyber Security (a unified national authority), Singapore's Cyber Security Agency (licensing cybersecurity providers), and India's CERT-In (mandatory 6-hour incident reporting) provide governance models in this area. Regional cooperation—through the AU, RECs, and bilateral partnerships—is essential for sharing threat intelligence and capacity building, as exemplified by the Organization of American States' experience.

10.7 National Financing Strategies

Strategies and efforts should be backed by sufficient financing. The continent needs $130–170 billion each year for infrastructure projects, but it currently raises only $85–90 billion. Domestic public resources account for about 45% of current investments but are constrained by weak tax collection (17% of GDP, compared with 34% in OECD countries) and high debt levels. National financing strategies should diversify instruments and harness underutilized domestic capital.

10.7.1 Mobilizing Domestic Capital

African pension and sovereign wealth funds manage over $2.5 trillion but remain largely untapped for technology infrastructure. Regulatory reforms allowing 5–10% of pension assets to be invested in alternative assets could unlock $3–4 billion in Nigeria alone. Tax incentives modeled after South Africa's Section 12J, which generated over R10 billion in domestically raised venture capital, can expand the capital pool. Diaspora remittances—$95 billion annually—are another underused source. Countries must also strengthen domestic tax collection.

10.7.2 Structuring Public-Private Partnerships

Morocco's telecommunications liberalization—with clear rules, regulatory independence, universal service obligations, and competitive dynamics—has attracted $12 billion since 2000 and achieved 95% coverage. Tanzania's national ICT backbone PPP, in contrast, shows the risks of poor design: vague performance specifications, disputed payment obligations, and a lack of transparency. Countries should establish transparent PPP frameworks with performance-based contracts and enforceable dispute-resolution mechanisms; use blended finance, viability gap funding, and first-loss facilities for strategically important but commercially marginal investments (such as rural broadband, mini-grids, and agricultural technology platforms); and build project-preparation capacity to develop bankable pipelines.

Kenya's M-KOPA illustrates progressive de-risking: development grants funded product development; concessional loans supported working capital; guarantees lowered investor risk; and commercial capital followed once the model was validated—enabling scale to 3 million customers.

10.7.3 Expanding Venture Capital and Innovation Finance

African venture capital increased from nearly zero in 2010 to $6.5 billion in 2023, but 80% of deals and 85% of the capital are focused in Nigeria, Kenya, South Africa, and Egypt. Fintech accounts for 40%, while agritech (6%), healthtech (7%), edtech (5%), and cleantech remain underserved.

Countries should create startup-friendly regulatory environments with simplified registration processes, clear exit options, and robust intellectual property protection. They need to establish national venture funds or co-investment mechanisms targeting underserved sectors and regions. Promoting university-industry commercialization through technology transfer offices and revenue-sharing incentives is also essential.

Supporting second-tier innovation hubs outside capital cities is crucial as well. Examples like Kenya's Silicon Savannah, Lagos's tech scene, Morocco's technoparks across the country, Tunisia's El Gazala technopark, and Cape Town's ecosystem demonstrate that vibrant African tech ecosystems depend on network density—many interconnected actors, frequent interactions, and trusted collaboration—rather than on any single institution.

10.8 Embedding Inclusion in National Technology Strategies

The book highlights that inclusion is crucial for the sustainable adoption of new technologies in Africa. Women constitute only 18% of tech entrepreneurs, and 234 million fewer African women access the mobile Internet. Youth unemployment and underemployment affect 57% of African youth. Rural populations have Internet penetration rates that are 2.6 times lower than those in urban areas. Without deliberate efforts, digital transformation will worsen these disparities.

National strategies must incorporate inclusion as a core design principle across all areas: implementing affordability policies to maintain data costs below the 2% of GNI threshold (through infrastructure competition, spectrum reform, and demand-side subsidies for low-income groups); launching gender-responsive technology programs to close the mobile Internet gender gap with targeted access initiatives, STEM scholarships for women, safe digital spaces, and support for women tech entrepreneurs.

Countries should also develop youth employment programs that link digital skills training with market demand (Kenya's Ajira Digital has trained 650,000 youth; Ethiopia's 5 million coder intiative has trained and certified over 300,000 youth; Nigeria's 3MTT initiative aims for 3 million); and produce content and interfaces in local languages—considering Africa's 2,000+ languages, which create barriers when most software only supports English, French, or Arabic.

10.9. Adaptive Learning and Accountability

Technology strategies often fail not because of poor initial design but because they cannot adapt. Countries should set measurable annual targets linked to each milestone and publish standardized progress dashboards; conduct independent mid-term evaluations assessing not only outputs (such as infrastructure built and laws passed) but also outcomes (such as cost reductions, productivity gains, job creation, and service efficiency); embed peer-review mechanisms similar to the African Peer Review Mechanism for shared learning; allocate sufficient M&E budgets (3–5% of program costs); and protect evaluation processes from political interference. Kenya's quarterly learning reviews—where stakeholders evaluate progress and adjust strategies—show that learning loops can be institutionalized without losing momentum in implementation.

More importantly, Africa needs to develop well-resourced research and analysis beyond what is currently provided by regional institutions, think tanks like Research ICT Africa, and university departments. Africa should establish an independent technology research center modeled on the European Joint Research Centre (JRC) to consistently provide reliable scientific evidence, data analysis, and technical expertise to support policy-making across the continent.

10.10 Conclusion

Africa's technological transformation is ultimately a national effort. While continental strategies provide frameworks and regional integration achieves scale, it is national governments that are responsible for building infrastructure, training the workforce, regulating markets, mobilizing financing, and creating the conditions in which technology either supports development or deepens divides. The evidence from this book makes it clear what determines success: strong foundations in energy, connectivity, and data infrastructure; substantial investment in human capital at all levels; governance frameworks that balance promoting innovation with managing risks and protecting rights; diversified financing that mobilizes domestic capital alongside international resources; and deliberate inclusion to ensure women, youth, rural populations, and marginalized communities benefit from — rather than are excluded by — technological change.

The pathways described here serve as frameworks, not prescriptions. Rwanda's government-led approach might not suit Kenya's market-driven environment. Nigeria's larger size allows for strategies that smaller nations can't access. South Africa's middle-income status offers options that lower-income neighbors do not have. Insights from Asia, Latin America, Oceania, and other regions may not translate effectively in Africa.

However, common principles apply everywhere: comprehensive strategies guided by evidence; inclusive stakeholder involvement; diversified funding; thorough monitoring; and adaptive learning that treats strategy as a hypothesis to be tested and improved, not a fixed plan.

Ultimately, success requires a generational commitment. Building infrastructure takes years; developing human capital unfolds over generations; and strengthening institutions demands consistent effort across political cycles. The combination of a young demographic dividend, a mobile-first innovation culture, abundant renewable energy and mineral resources, and increasing governance capacity creates favorable conditions for transformation. Africa's path is to forge a unique course—one that leverages the continent's distinctive strengths, addresses its specific challenges, incorporates its ethical values, and positions its young population not just as consumers of imported innovation but as creators of solutions with global relevance. The technologies are available. Governance frameworks are emerging. The demographic window is open. All that remains is sustained, coordinated, and adequately funded national action aligned with African values, priorities, and aspirations.

References

African Development Bank. (2024). African economic outlook 2024: Technology and transformation. AfDB Publishing.

African Export-Import Bank. (2024). Pan-African Payment and Settlement System: Progress report. Afreximbank.

African Union. (2024). Digital transformation strategy for Africa: Implementation review. AU ICT Division.

African Union Development Agency-NEPAD. (2024). Programme for Infrastructure Development in Africa: Progress report. AUDA-NEPAD.

Convergence Finance. (2024). The state of blended finance 2024. Convergence Publishing.

East African Community. (2024). Digital economy blueprint 2024–2028. EAC Secretariat.

ECOWAS Commission. (2024). Regional digital transformation strategy. ECOWAS ICT Department.

Egyptian Ministry of Communications. (2024). National AI strategy: Progress report. Government of Egypt.

ILO. (2024). World Employment and Social Outlook: Africa. ILO Publishing.

ITU. (2024). Measuring digital development: Facts and figures. ITU Publications.

M-KOPA. (2024). Pay-as-you-go solar: Impact report. https://m-kopa.com

Moroccan ANRT. (2024). Telecommunications market report. ANRT Publications.

Partech Africa. (2024). 2023 Africa tech venture capital report. Partech Partners.

Rwanda Development Board. (2024). Digital transformation roadmap: Five-year review. Government of Rwanda.

SADC Secretariat. (2023). SADC Integrated Regional Electronic Settlement System: Annual report. SADC.

South African Department of Communications. (2024). National digital and future skills strategy. Government of South Africa.

UNCTAD. (2024). Digital regulation in Africa: Survey of practices. UNCTAD Technical Papers.
UNESCO. (2024). UNESCO science report 2024: Africa regional analysis. UNESCO Publishing.
United Nations. (2024). E-government survey 2024: Africa regional report. UN DESA.
World Bank. (2024). Private participation in infrastructure database: 2023 annual report. World Bank PPI.
World Economic Forum. (2024). Global Competitiveness Report 2024: Africa Technology Strategies. WEF Publishing.

Index

Made in the USA
Monee, IL
05 November 2026

67914360R00164